The Napa Wine Industry

The Napa Valley Wine Industry:

The Organization of Excellence

By

Ian Malcolm Taplin

Cambridge
Scholars
Publishing

The Napa Valley Wine Industry: The Organization of Excellence

By Ian Malcolm Taplin

This book first published 2021. The present binding first published 2021.

Cambridge Scholars Publishing

Lady Stephenson Library, Newcastle upon Tyne, NE6 2PA, UK

British Library Cataloguing in Publication Data
A catalogue record for this book is available from the British Library

Copyright © 2021 by Ian Malcolm Taplin

All rights for this book reserved. No part of this book may be reproduced, stored in a retrieval system, or transmitted, in any form or by any means, electronic, mechanical, photocopying, recording or otherwise, without the prior permission of the copyright owner.

ISBN (10): 1-5275-6971-3
ISBN (13): 978-1-5275-6971-3

To the memory of my parents,
without whose support and encouragement early in my life,
none of what follows would have been possible

TABLE OF CONTENTS

List of Illustrations viii

Preface x

Introduction 1

Chapter One... 24
Wine Origins: Culture, Markets and Institutions

Chapter Two... 46
Missionaries and Adventurers: The Early Years of California Wine

Chapter Three.. 72
Nunc est Bibendum: The 1930s–1960s

Chapter Four.. 94
Behold the Future, Forget the Past

Chapter Five 116
Now is the Time for Fine Wine

Chapter Six 143
Affirming the New Orthodoxy

Chapter Seven.. 170
Conclusion: The Odyssey Ends

Appendix 190

Bibliography 193

Index.. 199

LIST OF ILLUSTRATIONS

Fig. 1.1 Napa appellations and sub-appellations ix

Fig. 2.1 Kvevri 26

Fig. 2.2 Chartrons wine warehouse district, nineteenth-century Bordeaux ... 38

Fig. 3.1 California Mission 48

Fig. 3.2 Eshcol Winery ca. 1890s 58

Fig. 4.1 Winery ca. 1930s 76

Fig. 4.2 Winery ca. 1940s 82

Fig. 1.1. Napa appellations and sub-appellations

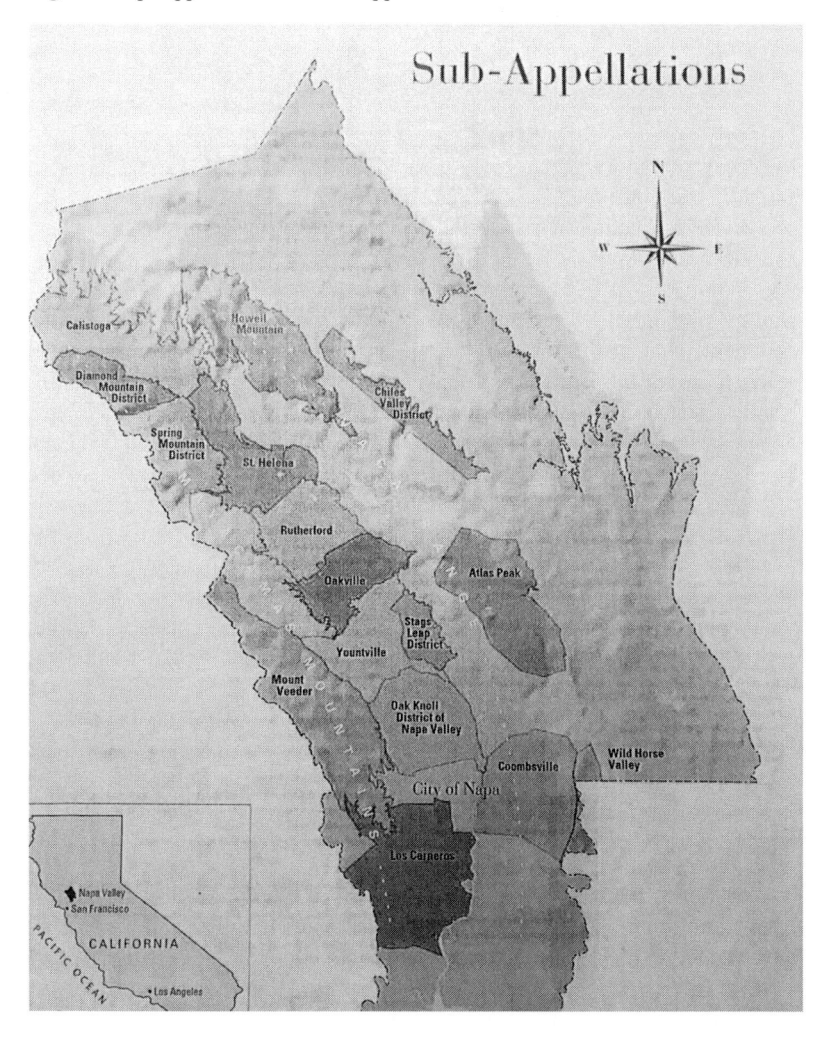

PREFACE

I started researching the Napa wine industry in 2005 in a somewhat serendipitous fashion. I was talking with a student about my new research project investigating the rise of the wine industry in North Carolina, where I'm currently based. I told her how I was intrigued by events in that state which had recently generated some initial successes after more than a century of failure. Specifically, I wanted to know how a market for wine emerged, what institutional forces were at play, and who the individuals responsible for this growth that started in the 1970s were. I told her that the history fascinated me, since many residents of North Carolina had traditionally demonstrated enthusiasm and efficiency in making alcohol, but had not fared well with wine. In the case of wine, it became obvious that the requisite technical skills and scientific knowledge were conspicuously absent in the past, and this would had not been rectified until late in the twentieth century. Having briefly relayed this story to her, the student simply replied "that sounds like Napa. You should talk with my dad, as he runs a famous winery there." And that is exactly what I did, contacting him and journeying to Napa to discuss the myriad issues that had framed the wine industry in that region and California since the beginning of the nineteenth century. That person was the late Tom Shelton, and we had a number of fascinating conversations. He also introduced me to other winery owners and winemakers who were gracious enough to share their time talking with me. This book is the result of those early discussions and the numerous other conversations that ensued over the years.

Rather than provide a rich narrative history of Napa which other authors have admirably done, and whose work I have used in my research, I wanted to do something slightly different. Specifically I sought to place this history in a theoretical framework that would enable me to discern patterns and processes. How did a market for wine evolve when it did? Which individuals brought valuable resources to start wineries? What institutional forces molded its embryonic structure? Which political events constrained it for several decades? I attempted to trace dynamic capabilities amongst firms and how they engendered industry growth when becoming part of interactions that fostered collective organizational learning. From this, I could discern how informal knowledge and formal exchange mechanisms morphed into structured relationships between firms, eventually leading to differentiation, albeit in a cooperative arena that allowed crucial tacit information to be disseminated.

We are familiar with the economics of agglomeration and the beneficial effects of cluster density, but there appears to be a different dynamic at work here, especially after the 1960s when the industry re-emerged following Prohibition. Efficient market theories presume rational behavior and transparency that guide economic growth, but information asymmetries and very different resource backgrounds by industry entrants can distort such efficiencies. This was evidently the case in the history of the Napa wine industry. A tremendous amount of cooperation between firms was almost a pre-condition for the region's eventual success – a very different proposition for the presumption of competitive rivalries. What became clear was how a determined and purposeful effort by many different actors to organize wine production around goals of quality replaced earlier efforts to build the industry as a commodity product. But quality would only work if consumers demanded it, and it is only recently that a wine culture per se has emerged in the United States. Some Americans have always drunk fine wine, but generally that meant Old World wines. What persuaded them to shift their allegiances and try wine from their own country? The answer, after wineries were able to consistently produce fine wine, was to have important critics affirm its quality, thus removing some of the uncertainty that consumers faced when buying an ephemeral product. Increased discretionary incomes meant more consumers could afford the higher prices that Napa wineries were charging – itself a pricing strategy designed to confer quality and status. Like many positional goods, an expensive wine was a mark of status; the better Napa wines became, the higher the price charged for them. A high price signified quality, and limited availability translated into a luxury product.

Old World wines took centuries to accumulate the prestige that we now confer upon them, and they underwent numerous attempts to order production around classifications that formalized quality. Napa has managed to achieve this status in a much shorter time. It too has relied upon systemic activities. Key actors, notably since Prohibition, have purposefully organized the industry in ways that capitalized upon endogenous features as well as their own resource capabilities (capital, skills, networks etc.), in doing so charting a course toward excellence. Collectively, they have been able to create an exceptional product, structure the market to protect their brand identity, and manage demand from a consumer base slowly acquiring sophisticated oenological preferences. The fact that they have built a stellar reputation in a fairly short time is quite remarkable.

A large number of close friends, whose names are too numerous to mention, have shared their insights on wine with me, and from whose comments I have benefitted. Thank you. When I first went to Napa, Russ Weis and Doug Shafer proved invaluable as sources of information, and have continued to be people I turn to when I have questions. Both have a truly remarkable understanding of the industry in its various nuanced forms. In recent years, Delia Viader and Tor Kenward have similarly been most edifying. Bill Harlan read and commented on the entire manuscript, for which I owe him a huge debt of gratitude. As with the others, he would throw out ideas and push me to probe areas that I hadn't initially considered important.

In the North Carolina wine industry, I continue to benefit from a wealth of information provided by numerous winery owners who I have come to know over the years. In this group I am particularly grateful to Mark Friszolowski, Chuck Johnson, Alan Ward, Jay Raffaldini, and J. W. Ray – friends who have always been willing to share their repertoire of knowledge on wine production. To them and the others, my thanks. In Bordeaux, my friends and colleagues Stephane Ouvrard and Hervé Remaud at Kedge Business School where I have taught have always been inspirational to talk with. Their knowledge of the French wine industry is truly impressive, and I have learned much from them – along with enjoying many excellent bottles of wine!

Finally, thanks to Cindy, who has listened to more discussions about wine than she would probably have liked, but without whom I'm not sure I could have accomplished this book, and many other things.

Introduction

I've been researching and writing about wine for almost two decades – specifically, how segments of the industry are organized, how they have changed, and what sorts of external forces have an impact upon production and consumption. When people discover this interest of mine I'm no longer surprised that they ask me what my favorite wine is and whether I can recommend one to them. These are two questions that I often find difficult to answer. Not knowing much about a person and when and how much wine they drink renders any comment I make possibly inappropriate. I sometimes ask, is it for special occasions, or a picnic, or an evening meal? Do they drink alone or mainly with friends? Context can matter. Furthermore, what I like might not be what someone else prefers. Taste is personal, whether in clothes, cars, films, leisure, or wine. For example, many of my close friends love Barolo, the majestic Italian wine that, quite frankly, I just don't get! I've tried it many times, and each occasion elicits the same indifferent response. They seem shocked when I tell them this, since they assume that someone with such knowledge about the industry would not make such an audacious statement. They ask if it's because I don't like Nebbiolo (the signature grape of Barolo). When I say I generally do like it, their response is that I probably have just not had a really good Barolo (read – expensive!). They are genuinely puzzled, and no matter how much I demur to their obviously superior palates, several start on a mission to provide me with ever more expensive and aged vintages in the hope that I will eventually have a vinous epiphany and become a convert. It has yet to work, but I appreciate the friendly disdain that continues unabated.

I have another friend who thinks Rhone wines surpass any others on the planet and is generally dismissive of Napa Cabernets because he thinks them too rich, oaky, and generally overpriced (which they can be). I respect his knowledge, and always enjoy tasting what he proffers when we get together (he's also an excellent cook, so the epicurean delights perfectly complement the wine – a topic I shall return to shortly). He is vastly more knowledgeable than me about wine, and he can discern all sorts of flavors that I struggle to identify. Nonetheless, from time to time I'm brave enough to disagree with some of his ideas, and he graciously accepts my criticism, although he did take issue the other week when I referred to a rare southern

Italian wine that he cherished as tasting like a cherry smoothie! Such are the conundrums of sharing one's thoughts with unabashed honesty!

When it comes to recommending wines to others, similar problems surface. When a person asks if such and such is a good wine, my response to them is often to say, "do you like it?" If the answer is yes, then I suggest that such a wine clearly satisfies their tastes, and the question of whether or not it is good is superfluous. As long as it is not flawed (and most people will recognize this immediately), then subjective perceptions should not be trivialized, and they should continue to enjoy it. By all means I encourage them to experiment with other wines that might be similar and thus expand their horizon, but to also be confident in what they like rather than having their taste dictated by others. For reassurance, I remind them that in clothing we have our own style, and while it is certainly influenced by others, ultimately it is what we feel most comfortable in. Do we always look good in what we wear? Possibly not. But if we're comfortable with our attire, despite the possible lack of sartorial elegance, we engage with the world regardless. Routines beget confidence.

The purpose of the above stories is to see wine as a diversified and differentiated product that can meet many needs and be subject to nuanced interpretations. It is a beverage that can be a simple drink that aids thirst as well as an iconic product that is quintessentially the epitome of excellence. Some people pride themselves on their knowledge and expertise in the latter, while others are merely in search of a decent complement to a meal. Others enjoy a glass of crisp white wine at the end of a hot summer day, or perhaps a bold red by the fire on a cold winter's evening. Their motivation and choice can vary dramatically. Many lack any pretension and willingly defer to others when seeking advice on wine. Often, they have discerned what they like and are confident buying it, even possibly extolling its virtues to their friends. In a recent visit to Costco I watched an elderly couple pick up a six pack of Apothic Red ($8.99 a bottle for a mass-produced red blend), and overheard the woman say how much she liked this wine, while her husband added that it is cheap. One aisle over, a young couple placed two bottles of Dom Perignon vintage Champagne ($149.99 a bottle) in their cart, commenting that it was a good buy and would be perfect for the celebration. They then added a six pack of a generic New Zealand Sauvignon Blanc ($9.99 a bottle) to their cart. I didn't hear what they said when picking the latter, but I must admit that my curiosity was piqued at such a combination of wines at opposite price points. Whatever the circumstances, these examples reveal how different people approach wine buying from varying perspectives, with a multiplicity of intentions, and, regardless of their wine knowledge, appear reasonably comfortable with their purchases.

In each of the above scenarios, it is individuals who make the decision, whether through self-evaluation or at the behest of critics and experts whose opinions can inform as well as tyrannize. But many of us continue to feel insecure when it comes to buying or choosing a wine. There's so much out there and, making a choice solely on the basis of a label (which in fact many are known to do) is somewhat fatuous, to say the least. For that reason, people look to outside experts to inform or validate a purchase. Such people can demystify wine, and if reliable can be a credible information source. For example, in an independently owned wine shop one might solicit recommendations from the owner whose opinion one trusts, since they have every incentive to understand the needs of what might be a repeat customer. There is such a shop in each of the towns that I live (Winston-Salem, NC and Cowes, Isle of Wight, England). In both cases, I have come to know the owners, and they make recommendations based upon what they know I like. They have been remarkably accurate in their suggestions, even when I might express an initial mild skepticism at a particular bottle.

For many people who seek out a bottle or bottles of wine, their destination is probably the local supermarket. They might know exactly what they are looking for, but if not, with less personalized attention, they are more likely to rely upon the descriptor tag that indicates "scores" awarded to a wine. If this influences their choice they are deferring again, but this time to an anonymous expert opinion.

From the above we can see that wine is a simultaneously complex and simple beverage that appeals to different people at different times, and in varying circumstances. It assumes an important role in culinary events, rituals, and celebrations, and has been an intrinsic part of Western culture for millennia. Stripped of pretension, it can constitute a mere commodity. In this case, it is something routinely drunk with meals by a large number of people – essentially a relatively generic product. But it can also possess quality characteristics that transcend the limitations of a mass-produced item. Here might be a wine of distinctive and refined quality, one that has an aura of craftlike perfection. In such elevated realms, it is the expression of unique characteristics and levels of skills that can eventually endow it with cult-like status. Then it becomes the apex of excellence and uniqueness, a testament to an obsession with perfection by those with the dedication and resources to sustain such a pursuit. Such products are often rare and generally expensive, drunk on special occasions, or on a regular basis by the very wealthy. To all extents and purposes, such wine is the quintessential luxury product.[1]

So, wine can be excellent, good, average, or indistinct, and possibly bad. History is replete with examples of each, especially the latter when, in early times, people struggled to understand the basic chemistry, how to mitigate flaws, and what sort of storage would best preserve the finished product. Wine did not travel well or keep for very long, so was generally consumed close to where it was made and quite soon after it was produced. Much of wine's long history fits precisely this description. And yet, even then, there have been some wines that have risen to the status of excellence, becoming well known, and generally located in regions that are now synonymous with fine wine (for example, Bordeaux, Burgundy, Tuscany, and Portugal with Port). Such reputations have been centuries in the making, with much trial and error by the producers before they learned which specific areas grow the best grapes, how to mitigate all the flaws, when to harvest, how to best ferment, what storage works best, and what container is the most resilient. Some of this was trial and error, but also the result of purposive strategies once quality benchmarks were recognized. Success was often the result of specialized knowledge, technical innovations, and individuals with resources such as time or capital, who had the patience and dedication to create a quality product. As these processes evolved, a de facto and eventually formal governance framework that ordered the marketplace became established. The latter is important because wine's global success is intertwined with the growth of a market which, together with increased trade and changing consumption patterns, is a crucial part of wine's evolving history.[2]

Wine as a Social and Cultural Product

Wine has an extensive and storied history. It has brought pleasure to many, but also pain to some. It has soothed the senses, heightened congeniality, and diminished woes. It was and continues to be a balm that calms nerves while exacerbating conviviality. It brings people together in communities, and for many cultures has become an integral part of culinary life, the perfect complement to a meal. It is a shared item, reinforcing bonds of friendship and kinship. Moreover, in societies where its longstanding history has endowed it with cultural significance, people don't necessarily talk about the wine but see it as a beverage that aids conversation. The Greeks famously used wines in their symposium (the Greek word *symposion* means "drinking together") where upper-class men gathered around a meal and discussed sundry items, some of elevated importance, while others were mere gossip.[3] While such activities were institutionalized as part of elite Greek life, wine consumption (presumably of lesser quality if not quantity) also occurred in normal homes and taverns by the common folk, who presumably also saw it as an aide to both digestion and conversation.

Rich and poor alike implicitly recognized wine's cultural and moral valence, and thus continued their bibulous ways without a second thought. They understood wine without intellectualizing it. It became a "taken for granted" beverage, infinitely preferable to water, which was often of dubious provenance and non-potable. Wine's place in such societies thus became firmly established, the result of centuries-old tradition and widespread usage.

Over the ages, wine also acquired a central role in rituals and celebratory feasts. It was a beverage that provided the focus for an event, the consummate symbolic act of drinking that reinforced one's commitment to a belief or ideology. In such instances, wine's role was more than a mere drink. It was a crucial part of an event, cherished for its ability to evoke complex feelings and sentiments. It sustained continuity in communities where rituals linked the past with the present. As such, it was often intricately linked with religion when consumption was equated with the attainment of spirituality. Such events elevated wine to a privileged status and prompted the introduction of quality mandates that would ensure the beverage possessed the integrity appropriate to the occasion. These concerns would come to noticeable fruition centuries later when medieval monks devoted part of their agricultural pursuits to grape growing and wine making (more on this later).

In countries with a more recent history of wine consumption, such entrenched cultural practices have been less common. And yet, when wine drinking did increase it gradually acquired sociocultural significance among certain groups. Often inspired by wine-drinking immigrants, wine consumption would eventually spread to broader sections of society, embraced by many for its aid to sociability. While product uncertainty among new consumers might still confuse, and purchasing confidence is often plagued by a lack of knowledge regarding wine varietals, this appears to have not deterred people from trying wine and developing an opinion on it. To give a coherent expression of their taste preferences other than sweet or dry, or red or white, might be difficult for many. But this doesn't detract from their enjoyment of a beverage that can satisfy a multiplicity of individual needs. For them, wine might lack the cultural imperatives. Yet this leaves them with an uncluttered imagination that can liberate their selection, and thus experimentation. They drink wine with friends and at meals, exchanging opinions, and generally the more they drink the more gregarious they become. In fact, the essence of sociability that is central to wine's consumption remains a crucial component of its enthusiastic embrace by new generations of drinkers.

Wine is thus a cultural product, its ubiquity framing centuries of consumption by ordinary people, nobles, and religious groups alike. But it is also an agricultural product with economic significance that has rendered many small communities viable. For people in such communities, the annual harvest was both a reward for their labors and the affirmation of community spirit. In evolving wine-producing areas, the annual calendar highlighted key agricultural events, and associated ceremonies further cemented the beverage's importance, both socially and economically. Production skills were noted, techniques learned and embellished upon, and certain individuals recognized for the quality of their product. Winemaking as an occupational specialty (*vignerons*, as the French termed such people) conferred status upon individuals who in the past had all too often relied on trial and error to make wine. Now they were becoming a constituent part of a network that progressively institutionalized a structure and organizational framework for such activity. Production and consumption became intertwined in narratives that highlighted geographical specificity, market formation, and eventually institutional coordination. Trade and commerce made wine profitable for some, while for others it provided a desirable product hitherto unavailable in areas where grapes could not be grown. Wine thus became an economic category that was crucial to many rural communities, further consolidating its social significance and cultural centrality, as well as prioritizing it as a value-added, differentiated product that transcended a mere commodity.

A Brief Historical Background

It is unambiguously clear when examining wine history that it has served multiple purposes: sociocultural, agricultural, commercial, religious, and even medicinal. As a physical product its origins are somewhat murky, and one can only speculate as to how and when our ancestors discovered that grapes left to overripen can ferment and produce a liquid that is pleasurable to imbibe. When they did realize this, in the absence of potable water, wine's role as a refreshment that aided digestion as well as provided a stimulus that eased the sorrows associated with the trial and tribulations of everyday life not surprisingly increased its utility. Presumably, there were many trial-and-error efforts to make wine, but little is known of these. What evidence exists follows the introduction of systematic production practices that led to wine becoming a commercially viable product. In other words, the practice of winemaking became institutionalized and somewhat formalized, indicating established procedures and a probable market for the product.[4]

Evidence that ancient Egyptians and Greeks used wine in religious rituals and increasingly in daily activities suggests that basic knowledge about viticulture was transferred across regions and cultures in the Eastern Mediterranean as far back as 6000 BC. But because it was expensive and often scarce, it was a beverage enjoyed by kings and elites, thus giving it a secular role to accompany its emerging religious and ritualistic significance. Wine was offered to the gods to seek favors, and its integral role in sacred life can be seen in the many depictions on temples and tombs where images of wine are pervasive.[5] In death, wine apparently provided solace – the afterlife might be fraught with uncertainty, but at least a pleasurable beverage was on hand.

Among the elite, its usage appeared to be associated with feasts and other important occasions. Eventually, it became a profitable agricultural commodity, benefiting from trade and increased consumption, which in turn drove more extensive planting of vineyards and flourishing commercial activity. While originally the preserve of the wealthy, by late Greek and then early Roman times its consumption had become more widespread among the general population. It retained its ritualistic importance, and presumably such wine was of the highest quality and the most expensive. Cheaper wine meanwhile was consumed daily by the rest of the population. Their habits, alongside religious rituals and elite consumers, gradually imbued wine with cultural significance as a commonplace beverage for many societies.

Wine gradually spread as a beverage of choice throughout Europe, often courtesy of Roman legions moving northward through Europe and planting vines to satisfy the needs of their soldiers. Such actions introduced wine to the cider and beer-drinking Gauls in what is now France. Centuries later, some of those vines came under the tutelage of monks who bestowed an enthusiasm, commitment, and increasingly systematic knowledge to perfect the beverage and raise the quality to standards that we are more familiar with today.

As the Greeks and Romans invoked the gods to legitimize their wine drinking and justify its importance as a beverage, medieval monks in France further enveloped wine with religious significance in their own winemaking activities. Religion would always imbue a legitimacy for a product, even one that had often developed negative connotations of revelry, debauchery, and degeneracy. But monastic orders provided a more rigorously ethereal and less sanguine context for the physical production of wine. They firmly established its centrality to daily rituals (the sacrament), as a complement to their daily meals, and an important part of their hospitality functions. Since the product was unambiguously reverential, the highest quality was essential, thus motivating the monks to work assiduously on perfecting their product. In this, they were aided by an organizational philosophy and long-term planning as they applied systematic methods of observation and calculation to discover how different plots of land produced different types of wine. Thus was born the notion of "terroir," and the idea that a multiplicity of factors combine to produce a distinctive type of wine.

This monastic legacy of scrupulous attention to detail and systematic experimentation plus enhanced technical knowledge laid the foundation for the production of high-quality wines, many of which have become notable brand identities. Here, and in other areas of France and Italy, wine's status as an everyday beverage remained, but the overall quality of the product was improving as people better understood ways of mitigating flaws as well as identifying what grows best and where.[6] Thus, markets became differentiated as wine became a beverage of distinction even when not designated as part of rituals. It retained its status as a commodity for everyday usage when not imbued with religious significance, but certain wines acquired the veneer of heightened respectability as a luxury product that elites and the wealthy could indulge in to enjoy and brag about.[7] A normal person might drink an inexpensive wine, possibly of their neighbor's making. But that wine lacks the quality and aura of a refined product. The wealthy, however, had discovered what their elite classical counterparts enjoyed millennia earlier – a beverage notable for its quality and expense, and one that few such as themselves could enjoy. In subsequent centuries, other regions in France and elsewhere in southern Europe became known for the excellence of their wine, while also producing vast quantities of cheap wine designed to slake the thirst of the average person. These issues are all discussed in greater detail in the next chapter.

What is a Quality Wine?

From this brief history one gains a sense of wine's integral role in culture, religion, and the daily life of ordinary people in past millennia. Wine became ubiquitous, imbued with cultural and commercial significance. It was both a commodity as well as a differentiated product to be savored on special occasions. Inevitably, some wines are better than others – that's natural, since different winemakers bring different skillsets, and different areas (through their soil, climate, topography, etc.) have the potential to produce better wines. But how can one tell what is a really good wine? Are there some objective measures against which one can evaluate a wine? And if so, is this something that the average wine drinker can detect? Presumably, the more wine one consumes the more discerning one becomes, although one is still at the mercy of one's memory. Inevitably, there have been many bad wines – ones that are flawed as a result of bad production methods. But if one discounts these, how much variation in quality exists within the bulk of wines, and who determines if a wine is of exceptional quality? Inevitably, pricing, classifications, and branding would provide answers to these questions.

From the Middle Ages onward, special wines were noted for their transcendent quality and also higher price. Consuming them proffered an aura of sophistication on the drinker because they had the financial means to acquire it. At the same time, some producers were recognized for their skills as *vignerons*, and certain regions became renowned for the distinctiveness of their wines. For example, Bordeaux producer Arnaud de Pontac marketed his wine (named *Château Haute Brion*) in London taverns during the mid-seventeenth century, being one of the first to assign a designation of place on the label.[8] Samuel Pepys commented favorably on this wine in his journals, referring to it as *Ho Bryn* (sic). Half a century later, the *London Gazette* commented on the quality (and high price) of other Bordeaux wines – Lafitt (*Lafitte Rothschild*), Margouz (*Château Margaux*), and *La Tour*. Clearly, quality wines were being produced and recognized as such by discerning consumers. Such wines commanded a high price and a regional identity – the genesis of subsequent brand development.[9]

But aside from price, exactly how do we evaluate wine when individual tastes vary and the circumstances under which we drink it can be very different? Do different people tasting the same wine discern the same flavors? Are some people better at identifying the unique characteristics of the wine, and thus the possible markers of finesse and excellence, than others? For the everyday drinker, does any of this really matter if all one seeks is a crisp white to pair with your oysters, or a robust, full-bodied red to accompany your steak? Finally, what is it that we as individuals like in wine?

When searching for a wine one appreciates, one is merely ambling down an oenological road littered with past experiences – some far more rewarding and enjoyable than others. One's taste changes, circumstances vary, opinions formulated alone are dramatically revised when with another whose authority is deemed more significant. No matter how confident one is in the ability to determine quality in a bottle, one can be influenced by external factors in ways that render any objectivity an elusive goal. A bottle of wine consumed at a picnic in a meadow on a beautiful spring day with the person of one's dreams, surrounded by rolling hills, gentle breezes, butterflies, and birds hovering overhead – everything evoking an Arcadian respite that soothes the senses and calms the nerves. How can the wine be anything but perfect? But that same wine drunk in a noisy restaurant, with boisterous table neighbors, unpleasant company, a searing headache, and the pervasive desire to escape to the tranquility of one's own domestic refuge – a flawed wine, perhaps, or an experience that significantly alters the perceived quality? Conceivably, this is a wine that would not be eagerly sought out on another occasion. And yet, ultimately it is still the same wine.

So perhaps it is not the wine that changes but the mood of the person drinking it. If such is the case then does this not render quality evaluations even more difficult? Or are there wines that can transcend these idiosyncrasies? If so, how do we know what inherent features signify such a wine rather than being determined by extraneous forces? Since individuals are not always confident in their own tasting, I suppose we can fall back on the wisdom of "gatekeepers" to authenticate the product. In recent years, critics such as Robert Parker and the writers at *Decanter* and *Wine Spectator*[10] have introduced supposedly objective measures of quality designed to capture what for many are intangible aspects of an ephemeral product. These include narrative accounts of flavor profiles with reference to fruitiness, floral, and other natural aroma descriptors, as well as comments on the wine's relative complexity and balance plus the introduction of a quantitative score (usually out of a hundred). For many, such critics perform a valuable role in breaking down information asymmetries. Others however might feel this merely obfuscates rather than edifies. Given consumer insecurity about buying wine, such guidance simplifies everything, reducing all of the imponderables and subtle nuances to a basic numerical score that they are familiar with from their school days. But does a score of one hundred mean a perfect wine? And if so, what do we mean by perfection? Yes, I agree that we can assign discrete accolades to a series of supposedly measurable traits and arrive at an overall assessment. Yet, can it be truly objective?

Another problem is that, once purchased, wine is consumed and thus disappears, except in the memory of the drinker. This ephemerality renders judgment even more confusing, because now one has to recall how the product tasted, and memories are notoriously inaccurate. We might remember a fine wine, but can we ever recall the exact taste? Would we be able to definitively say that the same wine drunk weeks or months afterward tasted exactly the same? Again, the situational effects and pervasive subjectivity all have a bearing on this. The philosopher Cain Todd questions whether wine is an everyday, ordinary, physical object, or an imaginative, interpretive, experiential one.[11] In other words, by giving greater agency to the drinker, are we rendering objective evaluations less meaningful? A flawed wine notwithstanding, what exactly is it that we appreciate when we taste a wine?

This brings us back to the essentialism of wine as a social phenomenon consumed at meals or with friends. From its origins, wine has been a social beverage, an aid to community spirit, a drink to foster congeniality, or an important component of rituals. This leaves me agreeing with Roger Scruton and his writing on wine from a philosopher's perspective. "A good wine," he says, "should always be accompanied by a good topic and the topic should be pursued around the table with the wine."[12] In other words, wine is not merely a good complement to food, but an even more crucial accompaniment to thought. The more elevated the conversation, the more salient the quality of the wine becomes. An extravagant claim, perhaps, but certainly one that resonates with what generations of winemakers have implicitly understood and embraced. Perhaps this is why there have always been producers who passionately strive to make a fine wine – the perfect expression of their capabilities as well as a product for the ages, and something that will embellish both the culinary and cerebral components of dining. They strive for perfection, recognizing that it is precisely an elusive goal and the challenge is inherent in the seeking – the journey is what motivates as much as the destination. They seek to make a product that corresponds to their own notion of beauty rather than one that has been imposed by others. As a consequence, there will continue to be great wines, well crafted, impressive, and redolent of all the characteristics that oenophiles sanctify. The rectitude of the critics as official arbiters of greatness will inevitably bring attention to such wines and their makers. But there are probably many unsung heroes in this journey who deserve a place in the pantheon, just as there are those who consistently produce a product that satisfies the tastes of many and in whose eyes perfection is less a transient phenomenon than the gratification at that particular moment and place in time.

Organizing a Wine Market

In the above discussion I endeavored to explore the vicissitudes of taste as well as the varied circumstances under which wine is produced and consumed. People have made and drunk wine for thousands of years, and while we do not entirely know what the quality was like in early times, we can appreciate that increased discernment among some wine consumers came as a result of improved wine quality by producers. This followed improved technical skills and knowledge transfer, and in turn presaged a demand for better-quality wine as people came to understand what was possible, and presumably had the financial wherewithal to pay for a quality product. In other words, an incipient market was organized around makers, sellers, distributors, and buyers of wine, and it is to such market activities that I now turn my attention.

Understanding the complex relationship between the supply and demand of wine necessitates an analysis of markets, specifically how they were created and structured, and how they subsequently evolved. In this next section I first discuss the general principles of market organization, and then discern how various agents might typically shape and structure exchanges in ways that engender efficiency. To understand firm, sector, and industry growth it is useful to contextualize the activities of individuals, institutions, and broader political economic forces within an organizational framework. Specifically, how can one account for the emergence of a market space that encouraged product differentiation and new entrants? I speculate that, for an evolving industry or sector to be successful, firms must be able to develop dynamic capabilities in conjunction with technological innovation, plus requisite resources (suitable injections of social and financial capital). How this pattern occurs is crucial in understanding market evolution as the product of purposive behavior rather than abstract deus ex machina forces that somehow emerge to reconcile disequilibrium.

If one conceptualizes markets as vehicles for organizing the exchange of goods or services, with the presumption that optimal levels of equilibrium are attained and thus efficiency maximized, can one assume that the interaction between supply-and-demand forces provides a nominal structure? This is the standard neoclassical view. But markets evolve gradually and often without initial coherence, save that of providing a semblance of structure for the exchange of goods. It is this gradually changing structure that eventually provides coordination and order that facilitates the production, distribution, and consumption of goods.[13] Typically, the structure emerges once a viable network of individuals endows consistency to exchanges. Such networks consist of actors whose evaluations of other actors provide legitimacy for market exchange or render it unviable. In other words, their willingness to invest time, money, and reputation is a crucial component for an emerging market's legitimacy and eventual success. When this occurs, consistency begets predictability, thus minimizing mutability and consolidating future market relationships.

The problem, however, is that knowledge is often imperfect and information asymmetric, and trust therefore becomes a crucial intermediary in such interactions. The attainment of trust is fraught with uncertainty and often misinformation that can exist if relations between firms are too competitive. Yet, if firms are to overcome the liability of newness, it is often in their interest to embrace the increased repetition and density of interactions. This is particularly the case with firms in an emerging sector that are seeking operating efficiencies that allow them entry into the marketplace. If they recognize the optimal benefits of engaging in interactions, this engenders trust and encourages the further exchange of both technical and tacit knowledge[14] – the latter being of particular importance in embryonic industry sectors where formal knowledge and practices are underdeveloped. The merging of informal knowledge and formal exchange mechanisms that eventually acquire semi-legal status confirms the salience of social relationships as an implicit structuring principal in market transactions. This is essentially what underlines sociologist Neil Fligstein's claim for the architecture of markets as institutionally embedded.[15]

Through mutually-agreed interaction and occasional actual resource sharing, firms develop capabilities that enhance economic routines. The resources that further sustain such routines, increasingly replicable by industry newcomers, generate collective organizational learning, leading to eventual industry-sector growth. Firms might subsequently differentiate themselves through their strategic values or capabilities and respective resources that generate value. But as Mathews argues, in the aggregate, their dynamic capabilities are part of an ongoing process whereby interaction between firms is the source of innovative capacity and economic learning.[16] In other words, firms utilize individual resources to leverage interaction and maintain routines to maximize economic efficiency, but draw from the clustered pool of other firms the organizational and operating dynamics that are diffused within such a framework. Such market opportunity learning dynamics contribute to innovation, both at the individual firm level as well as industrywide.[17]

The economist Michael Best argues that innovation and sectoral transformation are driven by entrepreneurial firms when part of networked groups of collectively innovating enterprises.[18] While his argument is derived from studies of high-technology sectors, it nonetheless resonates with what others have found (including the growth of distinctive wine regions) when small firms leverage diffuse design and technological innovations that are geographically proximate.[19] Additionally, entrepreneurial stability is enhanced when procedural issues can be informally transferred and requisite knowledge dissemination somewhat routinized. Furthermore, face-to-face interactions facilitate mutual adjustments that in turn foster capability specialization, experimentation, and incremental innovation.[20] The more successful that firms become, the more attractive the sector is to newcomers whose own resources can enhance embedded ties and concatenate new knowledge sources. Napa's recent history is a testament to this process.

An extensive literature on the dynamics of firm clusters goes all the way back to Alfred Marshall's seminal work highlighting this tendency when it was published in 1890.[21] Subsequent studies exemplify the process in which new industry entrants capitalize upon informal networks and tacit information acquisition to sustain their early growth. The continued co-location of firms in a particular area (or cluster) engenders knowledge diffusion and sustains collective organizational learning, furthering the coherence of an evolving network structure that eventually assumes a de facto governance role. Inter-firm collaboration embeds individuals, firms, and industry services in an evolving collaborative project that stimulates further innovation and eventually sector growth.[22] Clustering facilitates information sharing and collaboration, at least in the industry's infancy. It also permits firms with different resource capabilities to minimize weaknesses and maximize the benefits of collective pools of knowledge. Admittedly, weaker resourced firms might continue to experience adverse efficiency problems that could and often do see them exit the industry. But aside from such natural mortality rates, the structural aspects of firm density are generally conducive to cluster growth if inter-organizational relationships provide at least a semblance of informal governance and order.

As sectors grow and markets become established, competition along product lines emerges, but, at least in the initial stages, norms of reciprocal obligation continue to bind actors together because of the mutually beneficial nature of knowledge and information sharing. This, and an increasing access to emerging pools of skilled workers in which knowledge transfer occurs with inter-firm worker mobility, further adds to the dynamic capabilities of extant firms. Innovation subsequently becomes embedded in an institutional framework that legitimizes the market structure and provides nominal governance to support order.

Early Lessons from France

Crucially, markets not only satisfy the coordination of transactions between organizations and via networks, but also solve problems of production, consumption, and distribution. By providing an ordered space that brings key actors together, they lend predictability to transactions with determined outcomes less vulnerable to chance. For example, early markets for wine in and from France, focusing initially on Bordeaux and Burgundy, were hindered by the difficulty of transporting the product beyond a certain distance, thus consigning production and consumption to a local area. With improved transportation and innovations in winemaking techniques that minimized spoilage, it was possible to reach more geographically distant markets. But this necessitated greater product awareness by buyers, which came after suppliers (merchants acting on behalf of producers) themselves acted as de facto guarantors of quality. And this was achieved by virtue of their own informal knowledge of the local wine markets, plus an emerging relationship with third-party individuals who assessed winery production and shared information on pricing plus quality attributes.[23] This reduced information asymmetry to levels that inspired confidence among buyers, who increasingly recognized brand values.

The above unfolded over centuries, but it was in the mid-nineteenth century that Bordeaux, through the official classification system of 1848, formalized what were often informal assessments of the quality and ranking of wineries. This cemented a market structure which would confer identities and social standing to key actors whose prominence in vertical rankings enabled them to maintain boundaries that protected their new legitimized status.[24] This has subsequently evolved and embraced other regions within the country with a similarly acknowledged sense of place and quality. How might this pattern be different in an emerging wine industry in a different country without the tradition and culture of winemaking and consumption?

Wine economist Denton Marks identifies four crucial aspects of market behavior as it applies to wine, which are particularly salient for a new industry.[25] First is the level of information available to suppliers and buyers. Given the vicissitudes of climate, grape growing and winemaking have always been fraught with uncertainty, and this can be important in determining product quality. Second, market size and purchasing power can shape interactions – small producers might lack bargaining power and wealthy customers can influence market direction by their buying preferences. Third is the degree of competition and how this influences the market tendency to produce what consumers want. Finally, how efficient are markets in maximizing the performance of individual firms? In other words, without some regulatory framework will best-market outcomes always be achieved?

Clearly, knowledge in wine markets is imperfect, given the uncertainties listed above. Hence, issues such as pricing and quality are subject to a wide array of evaluations. Production cost variability might be less salient than in the past given improved technical knowledge, but, in an industry where one has only one chance a year to get it right, this is not something to be dismissed lightly. Since wine is not a homogenous product and quality varies considerably, it is imperative to gauge appropriate mechanisms that allow informed assessments. Classification systems can be useful in helping people understand and identify wine, and as sociologist Wei Zhao has argued, they are a core part of a regulatory system and help structure markets.[26] Such systems typically group wines according to some form of geographical or varietal identity. French wines are classified vertically according to regional appellation, whereas American wines are horizontally classified according to grape variety. In subsequent studies, he shows how such classifications are important for individual firm-identity construction, especially when it can be leveraged to raise prices if status indicators coalesce around reputational markers.[27] This allays some concerns over lack of transparency, although many classifications are in fact socially constructed and politically manipulated, and hence reward asymmetrical power structures and possibly distort objective measures. Nonetheless, they do provide a measure of market organization that can be useful for consumers in making purchasing decisions.

In sum, market efficiency is tempered by product ephemerality, significant differences in consumer motivation (drink now or lay down the wine), wine as an experience (as opposed to a search) good (you have to buy it and drink it to fully know what it is like), varying levels of producer capability (and knowledge), plus a complicated regulatory environment. All these add to the complexity, unpredictability, and sometimes lack of transparency that structure market behavior.

Outline of the Book

Perhaps the best way to unravel this complicated set of actions is to examine how the wine industry emerged and discuss the ways in which it served parallel markets (quality versus quantity wines), with certain products attaining status by virtue of their quality (or the perception thereof), and how producers organized the market around differentiated products. What can the Greeks, Romans, Burgundian monks, and Bordeaux burghers tell us about market forces and the institutionalization of a wine culture? Chapter one examines their story, and analyses how markets were created around product differentiation, trade, and changing consumer demand.

We then move to the New World, where in the past century regions have been recognized as sites where wines of the highest quality have been produced, often amid cultures where wine consumption was far less pervasive. In a shorter period, eschewing overt references to history, tradition, and culture and often embracing a more technologically mediated approach to winemaking, wineries in the New World developed requisite competencies and capabilities often around innovative techniques combined with extant practices. In many cases, they have been able to experiment and take advantage of favorable climatic growing conditions that suit certain types of wines and their flavor profile, and also embrace a more scientific approach to various stages on the production channel.

One such place is Napa, California, where over a century ago would-be winegrowers were still trying to determine what varietals best suited the region's soils and climates. They then saw their industry effectively shut down for decades following Prohibition and struggle for several more decades after the Second World War, before quality practices were more firmly institutionalized. Skill and knowledge levels became more widespread following the co-location of firms, and eventually foreign recognition of the area as a producer of fine wines was attained. Only then did a significant market for such wines become established as Americans increasingly took to this homegrown beverage and wine consumption increased.

In the ensuing chapters I examine how wine's place as a relatively simple beverage – introduced by missionaries into California several hundred years ago and initially used in religious rituals before becoming secularized – gradually transformed into an often-prized brand that in some instances has acquired iconic status. Through decades of trial and experimentation, grape growers gradually understood what varietals to plant, and eventually what consumers preferred. It is a story that embraces this evolving understanding of place and what the French call "terroir," as well as the science and technology that facilitated a rational and purposive approach to winemaking. Relatively free of regulatory frameworks that often constrained Old World growers, winemakers in California could experiment more freely than their European counterparts, and in one notable region – Napa – produced wines that are internationally recognized for their quality and style. This book is that story, one in which key individuals created a market for a hitherto undifferentiated product, and organized production in ways that have fostered status attributes and quality markers generally ascribed to wines from regions with longstanding historical (and cultural) traditions. It examines how individuals came together, with varying levels of experience, expertise, and resources, and shaped the evolution of an embryonic industry. In doing so, they have been able to assiduously coordinate activities among growers, share tacit knowledge, and eventually create an institutional framework that helped organize market structure. The result has been, in the past few decades, a purposeful attempt by actors in the marketplace to organize the pursuit of quality in their finished product and elevate brand Napa to the pinnacles of oenological excellence. What had taken centuries to achieve in the Old World has been accomplished in the space of decades.

Utilizing the organizational framework outlined earlier, I discuss how dynamic firm capabilities developed in Napa over the past one hundred and fifty years. While the historical trajectory of wine production in Bordeaux was longer, it was also marked by often ad hoc innovations, political upheavals, and structural changes. In Napa, actors were more purposive in their actions, and in recent decades assembled a market architecture that would facilitate quality and reputation building for an industry without the legacy of tradition and culture to fall back on.

Of specific salience is the way individuals brought different resource sets in different periods, and in doing so shaped the evolution of the marketplace for wine away from an early obsession with quantity to one that prioritized production quality. They did this in the context of agricultural transformation in northern California, from predominantly fruit-and-nut growing to vineyards. Their efforts were occasionally stymied by extraneous events alongside an imperfect knowledge of basic production techniques. But through their persistence and subsequent technical innovations, disseminated through institutional actors, they were able to forge an industry structure and eventual governance systems that ordered the marketplace. In later decades this attracted new actors, many with vastly different resources and capabilities who were able to elevate quality benchmarks, improve the status of the sector, and develop an informal classification that embellished the reputation of the region's wine. Key actors have distinctively shaped the evolution of California's wine industry, but this is not just a story of great men's role in history. Instead, I examine how institutional forces were deployed to help order the growing industry and how evolving forms of governance specifically targeted complementarity between actors, rather than competition.

Chapter two examines the early history of grape growing in northern California and the experimentation with different varietals, the arrival of individuals with sufficient resources to invest in the industry, specialists who disseminated information crucial to best practices, the growth of demand for wine (both locally and nationally), key organizations in the distribution and sale of wine, and the growing recognition of the potential quality of Napa wine. This occurred from the early to late nineteenth century. During this time, I show how firms frequently struggled because they often lacked the requisite skillsets, alongside the limited institutional support and technical knowledge, in the early decades. Not only did this hamper production, but the demand for the finished product continued to be limited as many consumers of alcohol preferred widely available cheap liquor. The market for wine was still in its infancy, and when it did exist it was often located thousands of miles from production sites.

All of this gradually changed, and by the turn of the century there was increased demand and enhanced winemaking proficiency among key actors. However, the onset of Prohibition in the twentieth century and its impact on the industry inevitably proved disastrous and set progress back. Chapter three chronicles the difficulties the few existing wineries faced in the decades immediately after the Repeal of Prohibition. Wine was still a minority agricultural product in Napa during this time, with fruit-and-nut growing being the preferred and most profitable endeavor. But, gradually, new individuals, new ideas, and new networks emerged. In this chapter, I examine the key players who reshaped the industry starting in the 1940s, bringing skillsets and eventually financial resources to shift the focus to quality rather than bulk production. During this time, an incipient market identity was established. This is the next important phase where social capital was instrumental in forging growth and the creation of collaborative networks that facilitated tacit information sharing. It also documents the growing institutional support that further codified knowledge together with enhanced technical skills made possible through an increased exploration of a scientific approach to viticulture and oenology that came from local universities. Each of these promoted productive efficiency. They also complemented the changes in the demand for certain types of wine, a restructuring of the distribution networks, and national product market growth. It was during this period that a defined market space for wine production arose in Napa, and firm core capabilities focused increasingly on product-led rather than price-led competition. Efficient production systems were a result of experimentation to determine best practices, as well as ongoing technological innovation.

Events took a significant change following the success of Napa wine following the famous Paris Tasting in the 1970s, when attention on Napa as a wine region emerged. In chapter four I detail the arrival of new resource-rich individuals to the industry as winery owners, improved winemaker skills following enhanced technical training, and the dissemination of resulting practices following mobility among winemakers as they moved from firm to firm. Key players such as Robert Mondavi were instrumental in building on the incipient international identity for Napa when he partnered with Bordeaux winery owner Baron Phillippe Rothschild to create Opus One. Corporations began to show further interest in acquiring wineries, but this was tempered by less-than-desirable returns on investment.

Chapter five chronicles the increasing planting of Cabernet Sauvignon and other Bordeaux varietals in the 1980s, replacing the workhorses (Zinfandel, Petite Sirah, and Alicante Bouschet) of previous eras. Cabernet became viewed as the quintessential premium value-added grape that grew well in the Napa climate. As vineyards were replanted, more newcomers entered the industry and the valley became an increasingly desirable residential location for those seeking a sophisticated rural idyll. Such trends prompted serious discussions about land-use practices, whether development should go unchecked, and if the number of wineries should be limited. The policy measures that were introduced included significant land-use regulation and the creation of an agricultural preserve that mandated minimum lot sizes for new properties. Attention also focused on Napa's identity and, following the creation of the American Viticultural Association (AVA) designation for Napa and eventually its sub-regions, debate ensued as to what proportion of grapes in a wine should come from Napa if that name is on the label. Ultimately, such governance issues were a corollary to the tensions between old and new residents.

During this decade, significant wealthy newcomers entered the industry with passion, commitment, and capital resources that enabled them to invest heavily in creating wines of exceptional quality that could match French first growths. The Old World remained the ultimate benchmark, but newcomers were determined to propel themselves into this august body by developing the most exacting practices that might replicate such quality achievements. This is the era of the so-called cult wines inception – small-lot producers with high-priced wines that were limited in availability and most likely sold on allocation. They catered to a new group of wealthy consumers who were an elite part of a continuously growing American wine culture.

In terms of reputation building, this was also the period when powerful industry critics, whose evaluation systems conferred a numerical score for wine quality, could make (or break) a winery. The critics were important because they demystified wine for many consumers who had hitherto balked at purchasing an unknown product. As noted earlier, wine is an "experience" as opposed to a "search" good. The latter affords consumers immediate evaluation of a product, whereas an experience good requires an initial purchase to ascertain quality. Critics took much of the uncertainty from purchasing higher-priced wines, such as those from Napa. Respected critics' scores also generated legitimacy and eventually cult-like status for a small group of producers with consistently high scores, and whose business model was unambiguously based upon excellent quality, small production, and limited availability, along with a high price. Demand for such positional goods was driven by a growing group of wealthy consumers who began buying Napa wine, partly as status symbols, and more recently because for some it represents investment-grade collateral. Finally the institutional environment underwent change as restrictions on growth and conditions of farming were imposed, thus constraining new winery founding but also increasing the value of the existing wineries (restricting supply).

Chapters six and seven discuss some of the recent political issues that have further constrained growth (land and labor scarcity, environmental concerns, and increased tourism), the threat of fires, and the gradual shift from a cooperative network that framed collaboration and enabled firms to externalize some capabilities tasks to a more atomistic competitive arrangement. A constantly-changing regulatory environment renders uncertainty, as do changes in the power of buyers following distributor concentration. New technology has enabled many wineries to sell direct to consumer and thus bypass distributors, but this can entail additional staffing costs. Innovation and technical skills continue to advance production capabilities but remain an instrumental feature of individual firms rather than complementary networks. Differentiation continues in a limited fashion, but new firm entry is restricted. Finally, the legitimacy of the region that came through critic endorsement has resulted in increased product homogeneity and the convergence toward a distinctive Napa style of wine.

A macro-organizational structure that gave rise to a unique, quality product and a defined marketspace, albeit one in which individual differences existed, has now become operationally more constrained, with less inducement for innovation. I speculate on how this study contributes to extant social-science theory, specifically on market formation, entrepreneurial growth dynamics, and intra-firm network capabilities that reconcile local skillsets with external resources. This book is an attempt to tell this Napa story, unravelling a narrative of hope over experience as pioneers created a market for wine production, and in doing so organized an industry that would eventually gain the reputational legitimacy that their Old World counterparts took centuries to achieve.

Notes

[1] Ian M. Taplin, *The Evolution of Luxury* (New York: Routledge/Taylor and Francis, 2020).
[2] Rod Phillips, *9000 Years of Wine* (Vancouver: Whitecap Books, 2017).
[3] Ibid., 58.
[4] Ibid., 58; John Varriano, *Wine: A Cultural History* (London: Reaktion Books, 2010).
[5] See Varriano, *Wine: A Cultural History*, chapters one and two.
[6] Rod Phillips, *French Wine: A History* (Berkeley: University of California Press, 2016).
[7] For a wider discussion of wine as an important part of the luxury goods industry see Taplin, *The Evolution of Luxury*.
[8] Stephane Ouvrard, Herve Remaud, and Ian Malcolm Taplin, "The Bordeaux Classified Growth System," in *Accounting For Alcohol*, eds. Martin Quinn and João Oliveira (London: Routledge/Taylor and Francis, 2019), 210.
[9] Phillips, *French Wine*; Wei Zhao, "Social Categories, Classification Systems, and Determinants of Wine Price in the California and French Wine Industries," *Sociological Perspectives* 51, no. 1 (2008): 163–99.
[10] There are many wine magazines, but these two have the gravitas that endows them with an authoritative viewpoint acknowledged by many.
[11] Cain Todd, *The Philosophy of Wine* (Montreal: McGill-Queen's University Press, 2010), 75.
[12] Roger Scruton, *I Drink Therefore I Am* (London: Bloomsbury, 2009), 3.
[13] Patrik Aspers, *Markets* (Cambridge: Polity Press, 2011), 39.
[14] Michael Storper and Anthony J. Venables, "Buzz: Face to Face Contact and the Urban Economy," *Journal of Economic Geography* 4, no. 4 (2004): 351–70.
[15] Neil Fligstein, *The Architecture of Markets* (Princeton: Princeton University Press, 2001).
[16] John A. Mathews, "Competitive Dynamics and Economic Learning: an Extended Resource-based View," *Industrial and Corporate Change* 12, no. 1 (2003): 117.

[17] The iterative nature between technological and market dynamics was central to Edith Penrose's theory of firm growth. See Edith Penrose, *Theory of the Growth of the Firm* (Oxford: Blackwell Press, 1959).
[18] Michael Best, *How Growth Really Happens* (Princeton: Princeton University Press, 2018), 114.
[19] Stefano Breschi, and Francesco Lissoni, "Knowledge Spillovers and Local Innovation Systems: A Critical Survey," *Industrial and Corporate Change* 10, no. 4 (2001): 975–1005; G. Page West III and Terry W. Noel, "The Impact of Knowledge Resources on New Venture Performance," *Journal of Small Business Management* 47, no. 1 (2009): 1–22; Ian M. Taplin, "Network Structure and Knowledge Transfer in Cluster Evolution," *International Journal of Organizational Analysis* 19, no. 2 (2011): 127–45.
[20] Best, *How Growth Really Happens.*
[21] Alfred Marshall, *Principles of Economics* (London: MacMillan, 1890). See also Peter Maskell, "Towards a Knowledge-based Theory of the Geographical Cluster," *Industrial and Corporate Change* 10, no. 4 (2001): 921–43.
[22] Udo Staber, "The Structure of Networks in Industrial Districts," *International Journal of Urban and Regional Research* 25, no. 3 (2001): 537–52.
[23] Pierre Marie Chauvin, "Globalization and Reputation Dynamics: The Case of Bordeaux Wines," in *The Globalization of Wine*, eds. David Inglis and Anna-Mari Almila (London, Bloomsbury, 2019), 103–14; Pierre Marie Chauvin, "Architecture des prix et morphologie sociale du marché. Les cases des Grands Crus de Bordeaux," *Revue Française de Sociologie* 52, no. 2 (2011): 277–309.
[24] Wei Zhao, "Understanding Classifications: Empirical Evidence from the American and French Wine Industries," *Poetics* 33 (2005): 179–200.
[25] Denton Marks, *Wine and Economics* (Cheltenham: Edward Elgar, 2015), 15–17.
[26] Zhao, "Understanding Classifications."
[27] Wei Zhao, "Market Institutions, Product Identities, and Valuation of California Premium Wines," *Sociological Quarterly* 50 (2009): 525–55; Wei Zhao and X. Zhou, "Status Inconsistency and Product Valuation in the California Wine Market," *Organization Science* 22, no. 6 (2011): 1435–48.

CHAPTER ONE

WINE ORIGINS:
CULTURE, MARKETS AND INSTITUTIONS

In the previous chapter I discussed why people drink wine and what influences their purchasing behavior. Most notably, we saw how wine is a sociocultural product rather than just a commercial one inasmuch as it has been integral to many aspects of community life. Sometimes, this was through religious rituals where wine sanctified ceremonies or was used as gift to placate the gods. However, it was its secular usage that solidified wine's centrality to daily routines either at the meal table or in a formal setting such as a tavern or bar. Here, wine might be the focus of a ceremonial occasion (weddings, anniversaries), a beverage that stimulates conversation, or a complement to a basic meal. For these and probably other reasons, humans have taken to this drink throughout the ages. This also explains why other people make wine since patently there was and is a market for it.

Alongside these varying consumption habits, markets for wine emerged, trade developed, and the commercial properties of wine became institutionalized. The more pervasive consumption became, the greater the likelihood that structures emerged to coordinate its commercial activities, especially in the social relations of trade that reflected prevailing geo-political situations and shaped market form. People, ideas, goods, and commodities spread along the ancient trade routes, and wine was certainly no exception.[1] Wine was, as sociologist David Inglis argues, a form of proto globalization in which the economic and cultural processes associated with grape growing and winemaking spread across borders.[2] We can see this among the Egyptians, Greeks, and then the Romans. The Greco-Roman world in particular elevated wine consumption, and this activity spread among inhabitants of other lands under their influence inspired by this behavior. Wine was becoming a commodity, but its economic importance is inseparable from its social presence. Crucially, it is important to recognize the persistence and interdependence of sociocultural attributes in the development of wine as a commercial product. As Phillips so succinctly states:

But climate and soil established only the geographical limits for viticulture and thereby a region's potential for producing wine grapes. The actual cultivation of grapes and the production of wine were human activities that reflected social, economic, and cultural decisions, and these were determined in turn by the economic and social value attributed to wine.[3]

Wine is thus a social construct, embodying the various facets of culture and values, as well as economic valence. Acknowledging this broad context is therefore crucial to any analysis of wine's evolving place in society, and is the analytic framework underlying the discussion in this chapter.

In order to comprehend how a simple agricultural product became such a valuable commodity, how markets for it were created and subsequently engendered trade, and how consumption patterns changed, we can look back at wine's gradual introduction in early societies. From its earliest known origins in around 6000 BC, to its use by Egyptian royalty and then its more widespread growth among the Greeks and Romans, one can discern basic patterns that have endured to this day. Its early importance in religious rituals continued through medieval times when Catholic monks dedicated part of their daily efforts to producing quality wines, establishing reputations that would eventually lead to the iconic Burgundy wines of today. Elsewhere in France, notably the Bordeaux region, wines were produced for domestic consumption and trade. In fact trade, first with England and then the Netherlands, would shape the evolution of this region's wine industry as technical improvements, sustained investment, and eventually an official ordering of wines according to their quality (and price) reinforced the status attributes of many producers there. Other Old World countries such as Portugal, Spain, and Italy saw their industries grow over the centuries, but rarely did they acquire the sustained resilience in which status attributes were accorded to many wines, as occurred in France.

In pursuing this historical narrative we can trace how winemaking skills and knowledge were transferred between cultures, a pattern that has been crucial for the product breaking free of its earliest regional specificity. The spread of tacit skills through migration would eventually lead to technical innovations that improved the quality of wine. But industry growth also occurred because wine proved to be a profitable agricultural endeavor which, alongside other notable value-added crops in the Eastern Mediterranean, provided a financial incentive for vineyard investment. As consumption increased and spread throughout Europe, markets developed to facilitate the growing production, distribution, and sale of wine. Since wine had acquired a cultural significance, whether in religious rituals or secular dining events, its marketability was enhanced.[4] Finally, as wine cultures became firmly established in more recent centuries, institutional frameworks and structures emerged to govern various facets of this industry. What started as a beverage for the wealthy, often under the guise of sacred rituals, became gradually democratized, before finally emerging as a differentiated marketplace where mass-produced wines co-exist with high-quality, scarce, and very expensive products.

Early History

Wine's origins are indeed quite murky. No one is exactly sure when wine was first produced, how it was discovered that grapes ferment, and who the first winemaking practitioners who applied horticultural skills to make it a drinkable beverage were. We can only speculate as to why people chose to consume wine when it became available, although there is some evidence that its early usage was medicinal.[5] For a product that has become so ubiquitous, it's curious that we know so little about how its usage started.

The earliest evidence of systematic efforts to make wine can be found in Georgia, circa 6000 BC.

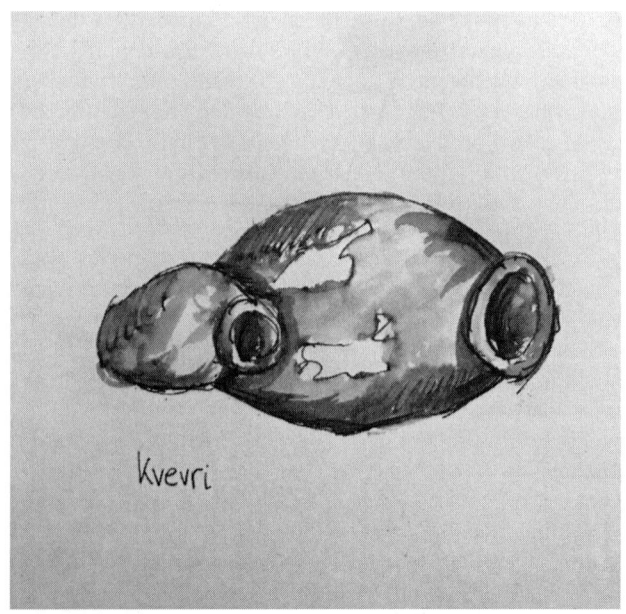

Fig. 2.1. Kvevri

Source: author's sketch

Storage vessels called kvevri have been discovered, which were used to hold unfermented grapes that were cultivated in the region. This suggests that a resolute production process to make wine existed since the grapes that were grown and harvested were subsequently crushed, and the container was buried to maintain a consistent temperature that allowed fermentation to proceed. The container was then firmly sealed and left for two years. What the wine tasted like we have no idea, although it was probably highly tannic given the evidence from recent efforts to replicate the early process. But what is important is evidence of an actual production process which implies some level of viticultural and vinicultural skills, a market for the product, and sufficient demand to necessitate the storage jars.

The Egyptians provide us with more systematic evidence of wine's presence, since wine jars have been found in tombs and burial chambers in the region of Upper Egypt. Markings on containers sometimes indicated vintage year, vintner's name, place of origin, and even comments on quality – an early recognition of the need to differentiate products for traders' usage.[6] Judging by the numerous inscriptions on temple walls, paintings, and engravings, wine appears to have played a significant role among the wealthy and religious leaders. It was apparently enjoyed in a communal settings, and religious imagery suggests a powerful role played by wine in sacred activities. Records indicate that religious groups cultivated vineyards of varying size, presumably using people with winemaking skillsets (and initially vines) possibly originating in the Middle East and the Anatolian region where there is some indication of an earlier wine culture existing.[7] We know little about these people and the extent to which their activities and roles were formalized. It's possible that religious groups learned these skills themselves, again with knowledge that came from other regions. What the wines tasted like and even the grape varietals are impossible to know for certain, but they did appear to be drunk regularly by the elite. There is, however, no evidence of wine consumption being widespread among the rest of the population, so one assumes it was an expensive product with limited accessibility, restricted to exclusive groups.

Our first evidence of a pervasive wine culture can be found in ancient Greece, where wine was apparently the most widely consumed beverage.[8] While it was typically diluted with water, it was nonetheless seen as a quintessentially masculine drink, with beer relegated to the female ranks. It was however no longer the preserve of wealthy and religious groups, and this relative democratization of the product helps explain the growth of a wine industry in Greece alongside the other two major Mediterranean agricultural products of grain and olives.[9] Demand for the beverage was increasing, and with it the supply. Greece became the center for the wine trade in the area, exporting the product around the Mediterranean. To solve the storage and transportation challenges, the Greeks used a similar-shaped container to the Georgian kvevri. These were called amphorae and were made of clay, with a design that made them easy to construct and use. Amphorae were widely used in Greek and then Roman times, their slim neck reducing the surface area of wine exposed to oxygen, their overall shape rendering shipping easy, and the handles at the neck facilitating ease of carrying and subsequently pouring. Examples of such containers have been found in many wrecks throughout the Mediterranean dating back to this period, suggesting a vigorous wine trade.

The storage of wine appears to have been a concern for these early producers inasmuch that wine could easily deteriorate under the wrong conditions, especially when exposed to oxygen. The use (and design) of amphorae suggests an awareness of such quality issues, and the necessary steps to mitigate flaws. Growing technical skills in winemaking also focused on making wine appealing to a variety of taste preferences. Again, we can only speculate as to the results of these innovations, but given that wine was increasingly part of not just meals but also important rituals, it is reasonable to assume that care was taken to ensure that it possessed the appropriate quality to match the occasion. As to the actual wine, from what we know choices were limited to red or white, sweet or dry, and strong or weak. Such limited differentiation might suggest a relatively unsophisticated palate, and perhaps limited winemaking skills. But, again, one can only speculate about taste preferences, although the heavy usage of wine at events that blended the secular and sacred suggests that quantity might frequently have negated any quality discernment. If the Greeks drank wine as an accompaniment to their discussions of truly serious questions, perhaps the demand for quality wine was intended to match the gravity of the topic.

Since interest in wine appears to have increased in Greece, it is worth pausing to consider salient aspects of Greek culture, and how these might have influenced the propensity to consume wine to the degree that they did, possibly regardless of the quality. What, if anything, can we deduce from everyday life in Greece that might shed light on the growth of consumption during this time? Life was hard in these early centuries, external threats omnipresent even at the best of times. This was clearly the case for the Greeks who ritualistically embraced uncertainty and institutionalized tragedy as a way of reminding themselves of the dangers of complacency.[10] There is perhaps no better way to witness this than in the rituals of drama that were an essential part of Greek life. They were designed to be a constant reminder of how things could easily go wrong, and wine's role was to ameliorate if not mitigate impending doom and catastrophe. Wine assumed a central part of such dramatic activities, presumably assuaging anxiety over impending calamity. Perhaps not surprisingly, the gods were seen as an instrument that tempered the futility of life. The pervasiveness of the cosmic order, however, didn't diminish uncertainty, but merely allayed the daily concerns.

It was to the gods that the Greeks looked, not just for explanations of sinister events, but also oenological governance as well as a formidable rationale for their daily consumption and occasional alcoholic excesses. The most famous for wine was Dionysus. He provided the raison d'être for earthly hedonism and invoked an aura of spiritual legitimacy for a wide range of activities that included agriculture, fertility, and general mayhem. In contrast to his cerebral half-brother Apollo, Dionysus glorified gratification and libidinous activities, and surely assuaged any guilt and self-recrimination that the Greeks might have harbored for their unruly behavior. Wanton debauchery might not have been explicitly condoned, but the boundaries for what might be deemed acceptable were certainly elastic.

For the Greeks and the Romans that followed, wine and the good life were clearly intermingled and pervasive but also resolutely sanctified by an otherworldly umbrella. By embracing divine intervention, one could seek meaning in events, search for solace as well as assuming a proactive stance against future tragedies. That was fine for existential questions and self-reflection, but what about more mundane deliberations? If there were any, they were probably overshadowed by the simple fact that the beverage was enjoyable, available, and normatively acceptable. Consequently, on a daily basis, people continued to imbibe a beverage with an ancient pedigree that enveloped secular habits with an aura of gustatory pleasure and widespread acceptability. We can only surmise that temperance was for many not a viable option or mandate.

In light of the sometimes-anecdotal evidence presented above, suffice to say that wine was central to many aspects of Greek life, and abstemious behavior not necessarily normative. Because of the apparent excesses that classical generations made central to their feasts, rituals, and celebrations, as well as even in daily life, concerns would eventually arise over the potentially destabilizing effects of wine drinking. Perhaps there could be too much of a good thing, especially if it might lead to corrosive antisocial behavior or be an impediment to work. Wine could bring people together, but also tear them apart if consumed in excess.

If the Greeks had symposiums in which men gathered to drink and discuss things, the Romans had banquets (*convivium*) that were far more inclusive class-wise, in which wine freely flowed and those in attendance overindulged in food. Topics under discussion were apparently more extensive, and presumably the intellectual tenor dissipated as the wine flowed. How much did they drink on a regular basis? As with the Egyptians, figurative imagery on Roman vessels suggests a high degree of wine-related revelry and debauchery, much of which was associated with the gods. Bacchus appears as an instigator in drunken behavior, as well as a more somber custodian of Roman military successes, or in the role of mediator between earth and the desired afterlife.[11] This intermingling of the sacred and secular, in which lines were blurred between routine dining events and those involving ritualistic behavior, sets the Romans apart from the Greeks. While the public face of wine consumption in Roman times was that of sacred intermediation, as seen in the many surviving images, one can deduce that it was also an everyday beverage privately, and thus not expensive. Romans of different classes drank wine, probably while dining or in a celebratory fashion with festivities. According to Cato, who gave "frequent and plentiful dinners," those in attendance came from many different walks of life.[12] It was apparent that wine was becoming less and less a drink for the elite, and there were sufficient supplies of mass-produced, lower-quality wine available.

Wine Production

We start to see evidence of a more systematic and informed approach to viticulture among the Romans, especially in attempts to address quality rather than quantity concerns. Grape cultivation was increasing, winemaking skills improving, and distribution and sales of wine indicate a vibrant marketplace. As wine historian Rod Philips has shown, one of the first documents on grape growing and winemaking appeared in AD 65 with a treatise by Columella.[13] This document set out in a remarkably thorough way the most beneficial grapes, the importance of soil conditions, and even rudimentary trellising forms. While debate flourished over many of his suggestions and the universal appropriateness of his claims, his work is nonetheless testament to a growing awareness of earlier viticultural limitations and a lack of precision. One of his most notable assertions was his focus on quality production. By paying attention to various facets of the production process and eschewing quantity goals, he showed how a medium-sized property growing good-quality grapes could produce wines for which one could charge a price premium, and thus be profitable. As with any agricultural endeavor, harvests varied, and it would be easy for a vineyard owner to lose money with a poor harvest. The quality approach mitigated this to a certain extent because, even with a small harvest, higher prices cushioned one's losses.

Given this differentiation in product characteristics, with higher prices afforded to wines of better quality, winemaking loses some of its mystery and winemakers their obduracy as economic opportunities become more viable and attractive. Not only could one make wine focusing on yield maximization for broad consumption, but it was now possible to make small quantities of better-quality wine and sell that to the discerning wealthy. Presumably, the Roman elite were willing to invest in such wines, and possibly these wines were reserved for more ceremonial functions as a recognition of their exceptionalism. For the rest of the population, however, their enthusiasm for cheap wine appeared to go undiminished, as indicated in the various edicts designed to stem what was seen as excessive consumption.

If Romans were apparently consuming large quantities of wine, this wasn't just to the detriment of basic civility. One senses that much wine was flowing when the Roman elite grew concerned that excessive consumption might undermine the effectiveness of the Roman military and their expansionist policies, and also lead to general unrest in society. An abundance of wine drinking was seen as potentially degrading young men and rendering them ill prepared for the rigors of military life. Unfortunately, the gods were patently of little help in rectifying this problem since they were often depicted as co-conspirators and less salutary examples of dignified behavior. How does one encourage moderation when societal norms appear to offer contradictory but sanctioned patterns?

Possibly to address these lingering concerns, various Roman Emperors such as Domitian in AD 95 attempted to ban the planting of new vines. If it was difficult to restrain demand, then reducing supply was seen as a viable solution. However, such edicts and policies were either not well implemented and practically enforced. Even though these actions could restrain competition and oversupply of wine, or even provide more land for grain and thus proving economically beneficial to the industry and society as a whole, they were nonetheless seen by many as an assault on their vinous way of life. The fact that this ban appeared to have been ignored and thus failed is an indication of how deeply rooted wine consumption had become in Roman society. The average citizen presumably did not take kindly to edicts from supposedly abstemious Emperors who were presumed to be less censorious when it came to their own consumption habits. As the population had taken to this beverage with an undiminished enthusiasm, wine had become an essential part of life, and attempts to regulate it destined to failure. It had become solidly rooted in Roman culture.

The Roman contribution to the spread of wine can be seen in the growing wine trade in areas where the Roman Empire was expanding, with Roman legions planting vines as they moved through Europe. Ostensibly to provide a supply of wine for the legions, such vineyards were instrumental in spreading wine culture throughout much of Western Europe as other populations took to the beverage. The beer and cider-drinking Gauls of what is now France soon came to appreciate wine following the Roman conquest, eventually recognizing it as a product that was not just socioculturally pleasurable but also commercially beneficial.

As one moves into the Middle Ages, the market for wine grew as demand increased alongside consumption. Yet again, the sacred played an important role in this early development since it was Christian monks who were instrumental in the next phase of oenological endeavors and put the pursuit of quality on a firm footing. It was also they who kept alive winemaking traditions, even though wine markets were depressed following the fall of the Roman Empire and during the Dark Ages.

Religion Once Again

Demand for wine increased in the Middle Ages because of urban growth and a merchant class whose increased wealth enabled them to partake of the luxuries of quality wine, hitherto restricted to the nobility and religious leaders. Wine, as Rod Phillips cogently argues, "had become entrenched as part of a wealthy lifestyle."[14] As more land came under cultivation with vines to meet this demand, trade prospered, as did the emerging merchant class that coordinated it and other commodities. Cheap wine was produced in growing quantities since it still remained part of subsistence agriculture in southern Europe. But quality wine as a value-added proposition was emerging in areas closer to large towns, such as Burgundy and Bordeaux in France and Tuscany in Italy. Here, religious orders, aristocratic landowners, and wealthy farmers along with prosperous merchants were planting vineyards – some for ascetic and sacred reasons, others according to a more secular profitability rationale.

The church was the most important institution in Medieval Europe, and its obligation to provide wine for the sacrament as well as a daily allocation for the monks proved an important impetus for them to develop winemaking skills. The virtual absence of secular records makes it difficult to discern whether monastic orders were solely responsible for engendering this next phase of winemaking and consumption, so our assessments are inevitably speculative. Conceivably, winemaking carried over from the ancient world via private enterprises, but unfortunately the extent to which that occurred is largely unknown. Instead, we tend to rely on monastic archives with examples such as the Abbey of Saint-Germaine-des-Près in Paris, which owned between three and four hundred hectares of vines, and produced 6,400 hectolitres of wine annually.[15] This wine was consumed by monks, and the remainder was sold.

The foremost religious orders in these activities were Benedictine and Cistercians monks.[16] They brought passion, conviction, and increasingly expertise to making wine, which became an integral part of not just their religious obligations but also their hospitality function. They accommodated travelers and cared for the poor as well as treated the sick. They were a focal part of rural life, powerful brokers of community affairs, and revered by many for their ascetic convictions.

The Cistercians were one of the first orders to turn viticulture into more of a commercial venture after founding their monastery at Cîteaux, Burgundy, in 1098. By continually expanding their holdings in the surrounding region, partly through the purchase of small properties but also as the recipients of land gifts, in 1336 they controlled fifty hectares. The amount of wine they produced clearly exceeded the demands of the Abbey, and while some might have been kept by peasants in return for helping them farm, other quantities were sold. Many such sales were local, but the growing recognition of their wines as being of high quality led to sales in urban areas such as Paris. The establishment of a regional market for quality wines, with individuals responsible for brokering and distributing the wine, augmented trade patterns and formalized exchange mechanisms around an incipient marketplace. The wealthy were driving sales with their enthusiasm for wine, and their growing numbers in urban areas signified a demand that many producers, presumably including monks, assiduously attempted to supply.

Systematizing Quality

While the issue of quality was presumably of concern to the ancients, in medieval times winemakers tackled problems with more direct intervention and greater authority. Availing themselves of extant knowledge alongside a willingness to experiment, they were much more systematic and innovative. The monks had the apparent luxury of experimentation in which grape varietals were planted, vineyard organization and management, and winemaking styles. They developed an understanding of what grape varietals grew best, and gradually developed significant winemaking skills that enabled them to make high-quality wines. They were deliberate in their experiments, trying various ways of cultivating vines, different soil times for planting, and vine orientation. They often grew different grape varietals side by side to determine which yielded the best-quality grape. If nothing else, they were very patient. Notwithstanding their embryonic commercial activities, since wine was ostensibly sacramental, it was incumbent on them to make the highest quality possible – to do otherwise would compromise their relationship with the Almighty. They sought perfection in whatever they did, and winemaking figured prominently in this sacred mandate. Theirs was also an endeavor for the ages, a multi-generational commitment to the pursuit of excellence. They were fortunate in having the luxury of experimentation without the fear of financial failure. Conceivably, other monastic agricultural activities cross-subsidized their winemaking operations to balance their finances, but we know that they were assiduous in their planning and determined to discover wines that would epitomize their sacred values and mission. Unhindered by short-term constraints, their long-term visions created the basis for Burgundy's subsequent measure of excellence and the site of some of the finest wines in the world, including what became Clos de Vougeot, which is now a prestigious Burgundian appellation.

Alongside the monasteries, wealthy aristocrats such as the Duke of Burgundy had extensive plantings, and by the late fourteenth century the Pinot Noir grape had acquired a reputation for the quality of wine it produced. Such wine was shipped to cities throughout northern Europe. Philip, the Duke of Burgundy in 1395, declared the harvest of that year "the best and most precious wines in the Kingdom of France for nourishing and sustaining human beings," thus cementing the reputation of the region's fine wines.[17] Comments such as this, although designed to enhance the value of his own vineyards, are nonetheless the beginnings of an eventual recognition of Burgundy as a region for the production of high-quality wines. It is difficult to ascertain how much wine and of what quality the majority of the population drank, and the data that refers to per-capita consumption is flawed. As Phillips has argued in his discussion of the limitations of such calculations for the medieval period, not everyone drank the same amount, and certainly many of the poorest could not afford wine.[18] Since water was free (although of dubious provenance), wine was a special treat for those who lived in winegrowing areas.

Wine production gradually spread through many regions in France, but it was in the Bordeaux area that some of the other most notable developments occurred. From the Middle Ages onward, in the region of Nouvelle Aquitaine, of which Bordeaux is the principal city, wine became an increasingly important commodity that was traded – principally to England, since the region was under the control of the English from the twelfth to the mid-fifteenth century. This growth saw the emergence of market-governance structures that helped coordinate transactions and smoothed the flow of requisite quality information. One of the best examples of this can be seen in the growing formal role of wine brokers (*courtiers*) in the Bordeaux region. Previously, wine was traded by merchants who marked up the price they paid and then sold on to wealthy elites, often in a different country. Now, *courtiers* acted as intermediaries, initially between wine estates and foreign merchants, but eventually with local merchants (*négociants*) in the evolving Bordeaux marketplace. They provided pricing, quality, and quantity information about wine that facilitated market transactions. Their role became institutionalized and professionalized, and they emerged as the first link in the broader distribution network.[19] This nascent market configuration proved useful as sales and trade increased, enhancing the economic efficiency and transparency for the key actors.

The Dutch, together with Flemish and Hanseatic traders, eventually replaced the English, and by the middle of the seventeenth century Bordeaux growers were making fuller-bodied and sometimes sweet wines that catered to the tastes of this new group of customers.[20] Furthermore, the Dutch used their extensive land reclamation expertise and assisted in the drainage of large swathes of land adjacent to the Garonne river near Bordeaux, thus opening up property that was suited to growing vinifera grapes. Wealthy merchants and aristocratic landowners developed this land into vineyards and applied new techniques that resulted in fuller-bodied wines that could last longer. The latter was crucial for international trade, as wine had frequently been subject to spoilage when traveling long distances.

During the next century, yields increased in response to a growing demand for inexpensive, poor-quality wine for the majority of the population. Yet, there was also an increased demand for lower-yield quality wines, driven in part by the urban middle classes in addition to wealthy merchants. Quality was improving as winemaking benefitted from continued technological innovation and a much better understanding of viticulture, including the area of yield management. Certain varietals produced consistently better-quality wines when yields were systematically reduced. Such wines could command a price premium, and demand for such wines continued to grow.

Bordeaux and Burgundy (plus the Champagne region) were acknowledged to have the best wine, and also the most expensive.

The logic behind this reputation building is not dissimilar to many luxury goods. Since people were willing to pay a lot for such wines and demand was high, ipso facto they must be of good quality. And, as Phillips notes, "at the higher end of the price range, it was wealthier and therefore more powerful and prestigious consumers who were prepared to pay more for wine, and these consumers were the arbiters of taste in wine."[21] Wine almost inevitably tastes better when you know the wealthy are buying and drinking it because they are presumed to be connoisseurs!

As with Burgundy, site-specific growing conditions were seen as an integral part of a winery's reputation, and thus the notion of the terroir developed.[22] Wineries would reference this as an explanatory but difficult-to-imitate factor in the quality of their wine – a justification and explanation that resonates to this day for certain well-known producers. It was during this time that some wineries attached their own name and place identity to their wine and marketed it accordingly, thus beginning the development of brand identity.

By the late eighteenth century the wine trade had become more organized and institutionalized around the activities of winery owners, brokers, merchants, and local authorities. The latter, the Jurade of Bordeaux, provided a structure that helped coordinate wine production and sales, judge wines, and promote the industry to wealthy individuals in the north.[23] Brokers continued in their importance during this period because their mutual trust with market members smoothed the efficiency of transactions. They had extensive knowledge of wine growers, the quality of their techniques, and the wine they produced, and their opinions were trusted by buyers as representing this information as objectively and honestly as possible.

Aside from such access to specialized knowledge, how could one determine the quality of a particular wine when so much came from a particular district? Previously, in Bordeaux there had been attempts at informally classifying various wines to separate the good from the mediocre and bad. However, as the wine market became institutionalized, the need for a more official ranking that could systematically classify wines according to quasi-objective criterion became increasingly necessary. This could eliminate confusion and avoid some of the exigencies of subjective judgment that plagued earlier informal categorizations. The result of these longstanding deliberations was the official 1855 classification which formalized hitherto informal assessments of wine quality by identifying and hierarchically ranking wine estates. Ostensibly the result of Emperor Napoleon III's request to Bordeaux merchants to select the best wines for the Paris Universal Exhibition of 1855, it largely built on extant valuations in which price was the principal determinant of quality. The higher the price, the better the presumed quality. It also used a territorial dimension (Medoc and Sauternes) that was a proxy nod to terroir, and structured estates into five classes (first, second, third, etc.). But it was also shaped by merchants, of whom the English (as one of the principal markets for fine Bordeaux wine) exercised disproportionate power over. Together with local merchants, they easily classified the estates they traded with, and apportioned highest rankings to those where the demand was greatest.[24] It's somewhat paradoxical that a simple logic – if there is high demand for a particular wine, then it must be good and thus expensive – pervaded much of the evaluation process, but that is how it emerged and became formalized.

The final result was a structured statutory space in which the owners of many existing estates whose wines commanded a high price were able to close entrance to new participants (whose wineries were not classified), and thus solidify their pre-eminent position. Even though the final classification was the accumulation of a longstanding system whereby public authorities, professionals, and amateurs determined rank on the basis of criteria such as terroir, quality, appellation, and grape varietal, what became privileged was price. The institutionalization of this process has endured, and remains fairly rigid to this day.

The owners of many of these estates were often wealthy businessmen who, together with a few aristocratic families, had bought properties and financed vineyard and winemaking investments. Ultimately, they consolidated their position in the Bordeaux marketplace. Unlike Burgundy, where estates were often quite small, many of these were large and thus afforded certain scale economies and seller power in the value chain. They had longstanding reputations based on historic overseas trading patterns, and were resolute in maximizing the legitimacy of such traditions. Often working in cooperation with *courtiers* and *négociants* (wine merchants), whose role had been formalized in the marketplace, they were able to exercise a considerable degree of control over prices. The classification system conferred significant reputational benefits, which had the effect of reducing some of the uncertainty inherent in wine production for top producers. That, and a

Fig. 2.2. Chartrons wine warehouse district, nineteenth-century Bordeaux

Source: author's sketch

regulatory framework that imposed output restrictions and varietal specifications, insulated them from much market uncertainty. As Ouvrard and Taplin argue:

> As the reputation of Bordeaux's wines increased, the demand for the best chateaux intensified while supply remained fairly stable because of production regulations. Vintage mattered because weather conditions affected output, both in quality and quantity. This put additional pressure on the operation of the wine market in Bordeaux. In good years demand far exceeded supply; bad vintages were correspondingly much harder to sell. The top wines, however, remained impervious to these fluctuations.[25]

The French Revolution had more dramatic consequences for Burgundy than Bordeaux because many of the former's owners were nobles, or, as noted earlier, under monastic control. When the state confiscated all church land and much of the nobility's, the vineyards were sold to pay off the monarchy's debts. Much of it was purchased under auction (to maximize the returns), and the estates were often broken up to further increase sales, leading to the fragmentation that characterizes Burgundy to this day. Local bourgeoisie and wealthy peasants acquired small parcels (*vignobles de paysans*), sometimes less than a hectare, although the better-known estates were less likely to be broken up and attained a higher price because wealthier individuals were willing to bid for them and were powerful enough to influence the auction proceedings. Such land transfer did not occur with the same ferocity in Bordeaux where there was less monastic land, and wine estates were more likely to already be in the hands of wealthy merchants (*vignobles de marchands*).

With their reputation for fine wines firmly cemented by the 1855 classification, Bordeaux estate owners continued to invest in improved viticultural techniques and also reduced yields because their wines commanded higher prices, thus negating any disposition toward increasing quantity. Many added the prefix "chateau" to their property because it spoke to a sometimes-imagined aristocratic origin, and lent a further degree of tradition and respectability to their wines. They assiduously managed their reputation, continued to build export markets, and were fiercely protective of their now-formalized status. Only one (Château Mouton Rothschild) managed to gain entry to the privileged first growth from the top of the Second, and that was not until 1973. This immutability has often been criticized as outdated and no longer a true reflection of actual quality distinctions among estates. Nonetheless, its persistence continues to drive sales, and demand for classified growth wines remains very high.

The area of St Emilion to the northeast of Bordeaux, referred to as the Right Bank (Medoc is known as the Left Bank), developed a similar but far less rigid classification in 1955, two years after the one for Graves. In both cases, estates were not ranked but merely classified (*premier grand cru*, *grand cru*, etc.), and identity was somewhat more fungible as up-and-coming estates could gain such a designation. It was an evolving designation and also based on a more systematic evaluation of the quality of the wines, derived from expert tastings and an attempt at objectivity. But, as with Bordeaux, price was also an important variable. Those with the highest prices were automatically deemed the top quality.

Since many estates were small they often lacked the branding power and reputation of the Medoc châteaux, nor did they necessarily have easy access to the networked structure of an intermediary relationship in the marketplace. With a few exceptions, they lacked the pricing power that their classified counterparts possessed, more likely "price takers" than "price makers" in their wine sales.[26] Reputation building has sometimes proved more difficult, especially for newer wineries, although they have managed to capitalize on the collective identification afforded by their classification status. Not surprisingly, status building is seen as a crucial component of pricing strategies, with enhanced agency most likely to go to longstanding incumbents in a region.

For the remainder of the nineteenth century, wine consumption generally increased – as did production, but with much volatility. Growing demand from urban areas in northern Europe for mass-market wines plus the consumption of cheap wine on a daily basis by the French working class encouraged further production of high-yield varietals.[27] The railways had integrated markets and lowered transportation costs, and living standards were gradually rising. To meet this demand for wine, vineyard work underwent a transformation following the increased use of day laborers as part of capitalist economic relations that were replacing vestiges of peasant labor. Much of this wine came from the south of France and then Algeria, where production by large companies increased from 1 million hectares in 1885 to 8.4 million by 1910.[28] Many of the newer wineries were large capital-intensive operations, but harvests still determined yields and therefore prices, with one good harvest every three years at times in the century.

As the mass market grew so did continued interest in quality wine by more discerning and wealthy individuals. Again, foreign demand was high, with the United States joining northern Europe as an important market. The result of these trends was an increasingly segmented market. Generally speaking, Bordeaux, Burgundy, and the Champagne region specialized in high-quality premium wines that were designed for export, while southern France and Algeria made bulk ordinary wines for mass consumption in the domestic market.[29]

Despite vineyard investments and improved technical winemaking skills, diseases continued to be a problem. Powdery mildew reduced yields, and downy mildew limited the alcohol strength and longevity of wine, both afflicting mass-produced and quality producers alike. But it was the onset of phylloxera in the mid to late 1800s that proved far more devastating to vineyards. The disease, caused by a tiny aphid that attacked and destroyed the root system, killed the vine. Eventually, it was responsible for the destruction of most vineyards in Europe, until it was recognized that grafting disease-resistant rootstocks from American vines eliminated the problem.[30]

Initial uncertainty over the causes of phylloxera, with many unsuccessful experimental cures attempted (including the flooding of vineyards to drown the aphid), was accompanied by opposition to importing American rootstocks since they might diminish the value and reputation of French vines. Viticultural knowledge had been steadily improving but was still woefully lacking in scientific rigor. The result of this void, alongside the reluctance to make the necessary changes when successful treatment of the disease was acknowledged and implemented, was considerable production losses and the collapse of many small *vignerons*. Because production was so variable, merchants intervened in the market by sourcing wine from different areas and blending them. This was a significant step toward organizing the market in ways that protected their position and role. In doing this, they were able to produce a standardized product that still satisfied consumer demands as well as continuing to work with quality producers to ensure that their wine could be guaranteed.[31] Their active involvement paralleled more government intervention in and regulation of the industry at the turn of the century.

By the early 1900s, relaxed licensing laws encouraged more wine consumption, and government efforts to reduce misrepresentation resulted in the establishment and designation of distinctive regions on the basis of the wine they produced (i.e. Bordeaux, Champagne, etc.) in 1908. The latter led to the establishment of rules regulating quality of wine, the types of grape varietals used, and yield restrictions. This was formalized in 1935 with the creation of the Appellation d'Origine Contrôlée (AOC) – a comprehensive framework to govern the industry as well as provide stability to what had been a tumultuous period at the turn of the century. Production had suffered from phylloxera in the late 1800s and then the First World War disrupted demand for all but the low-quality volume wine provided to the French troops. High mortality rates among men during the war further depressed the market. But it was earlier concern among the quality producers over fraud – by small *vignerons* over unfair competition from large producers, and by merchants eager to reassure their foreign customers of the integrity of the wine they were selling them – that led to pressure from below, encouraging the government to act. The result has been a structured marketplace with procedures designed to augment incumbent efficiency by managing the supply of wine. Its legitimacy has proved resilient since, with some modification, this regulatory system continues to this day.

The systematic transmission of scientific and technical viticultural and vinicultural knowledge occurred during the twentieth century with programs at French universities in key winemaking regions. This reduced much of the uncertainty that had plagued producers in the past, giving them more opportunities to ameliorate problems in the vineyard. Nonetheless, the distinctive attributes and explanatory power of the terroir remain the *sine qua non* for many producers, especially in the smaller plots in Burgundy, but there has been a growing recognition that technology and science can be huge resources for improving wine quality and consistency. While gaining access to such formalized knowledge is widely available, many winemakers still face the difficulty of embracing the new without relinquishing their hold on tradition and their primal linkage with the land. Most, however, recognize that poor harvests – which in the past resulted in significant problems with grape quality – could now be better managed. Annual variations still occur, but whereas a poor harvest previously resulted in inferior wine, the result would now be merely a lower yield. This was beneficial for elite producers, who saw an improved pricing stability since quality was less questionable.

In recent decades, France witnessed a significant decrease in per-capita wine consumption from one hundred liters per capita in the mid-1970s to less than fifty now – a result of younger people choosing alternative beverages as well as the more aggressive enforcement of drink-driving laws.[32] This has been particularly damaging for mass-produced wines because of their lower margins and volume. Because of this declining demand, public-policy regulations have introduced dramatic changes for this sector, from making wine into industrial alcohol to pulling up large numbers of vines and offering subsidies for affected producers. This has not affected the premium sector, however, especially since much of their market demand comes from overseas, including significant increases from China. If anything, such changes have pushed wine into a higher value-added category for the remaining producers, and focus even more heavily on export markets. But it does suggest a significant change in French wine culture, which, after centuries of variable growth, is clearly waning.

Conclusion

In this brief historical overview, I have attempted to identify key events that have shaped the emergence of a wine culture, market formation and the growth of trade, evolving production and consumption patterns, and the eventual systematic segmentation of the market into quality versus mass-produced, high-volume wine production. It is impossible to delineate the many intricacies involved in these processes, and others have done this most comprehensively and eloquently. Instead, my focus has been on the underlying patterns and trends, and how key actors have emerged to shape and eventually structure the market. Of the early times, we are limited in our detailed knowledge, often relying on imagery that is suggestive of behavioral patterns along with limited written testimonies. In the Middle Ages, our reliance on written records tells us much about the growth of quality production associated with monastic orders, but little of what small producers were doing, let alone what peasant farmers were drinking, and how much. Again, Bordeaux emerges as a distinctive area associated with quality wines, but much wine was produced in this area over the centuries that was presumably of inferior quality.

By examining markets and trade, one gains a better sense of how a supply-and-demand equilibrium evolved, but, as with any market operation, it tends to reflect the interests and power of key players. The fact that Bordeaux merchants were instrumental in organizing the market in the late nineteenth century to safeguard their own profits, and estate owners as well as small producers pushed the government to regulate the industry and adopt standard practices, suggests a slightly more inclusive system of influence. Nonetheless, it is the activity of key players in the supply side of the market (*chateâux*, *courtiers*, and *négocian*ts) that was instrumental in framing the operations of the Place de Bordeaux (the formal market for the sale of wine), which retained an aura of exclusivity that delimited who and what could be sold, and under specific terms. This type of organizational structure proved to be very resilient for the marketing of Bordeaux wines, especially the better-quality ones, since it managed restricted supply and thus manipulated demand. It was an evolving proactive attempt to organize the market that would frequently be plagued by dissent from those excluded from the upper echelons, as well as tensions between mass-produced versus quality producers.

While wine had been a luxury good with limited availability and a high price, it was eventually democratized, and market bifurcation occurred. Hitherto a product accessible only to the rich and those of religious orders, it gradually became a beverage for the common person. In doing so, it lost some of its mystery and distinctiveness, but this was eventually regained when better viticultural knowledge permitted producers to successfully focus on lower-yield quality grapes. Aside from the monastic orders, who had the luxury of being profitability averse and produced quality wines, wealthy merchants and some nobles brought resources that would eventually improve quality and consistency. Technical innovation and enhanced winemaking knowledge enabled those with adequate financial resources to make the necessary improvements. They combined a reverence for tradition and lineage, the centrality of the terroir, and a forward-looking embrace of scientific rigor. Their progressive attempts to demarcate their site specificity through labeling and branding enabled them to create a marketspace for a high-quality and expensive product, with limited supply, that would legitimize their identity and status reputation. The hierarchical classification systems that ensued merely formalized this position.

Similar trials and tribulations were associated with the wine industries of Italy and Spain (and to some extent Portugal, but there mainly with Port), each of which were able to enter the pantheon of oenological respectability. However, I have deliberately enunciated the French history in this chapter because Bordeaux and Burgundy became de facto benchmarks for Napa, their story resonating so closely with what aspiring winegrowers in Napa deemed essential for success.

I now turn my focus to the New World, where wine cultures were emerging – specifically California, where a combination of events presaged what would become an important area for wine. I will trace the development of the wine industry in California during the nineteenth century and the struggles for identity, quality, market formation, and appropriate grape-varietal selection that shaped the industry's growth. Within California, Napa would eventually emerge as a distinctive wine region when it was realized that certain varietals grew well and could produce potentially excellent-quality wines.

Notes

[1] David Inglis, "Wine Globalization: Longer Term Dynamics and Contemporary Patterns," in *The Globalization of Wine*, eds. David Inglis and Anna-Mari Almila (London: Bloomsbury, 2019), 23.
[2] Ibid., 22.
[3] Rod Phillips, *9000 Years of Wine* (Vancouver: Whitecap Books, 2017), 23.
[4] Ibid., 23–5.
[5] John Varriano, *Wine: A Cultural History* (London: Reaktion Books, 2010).
[6] Inglis, "Wine Globalization," 22.
[7] Ronald L. Gorny, "Viticulture in Ancient Anatolia," in *The Origins and Ancient History of Wine*, eds. P. E. McGovern, S. J. Fleming, and S. H. Katz (Luxembourg, 1996), 151–2.
[8] Varriano, *Wine: A Cultural History*, 23.
[9] Phillips, *9000 Years of Wine*, 44.
[10] Hal Brands and Charles Edel, *The Lessons of Tragedy* (New Haven: Yale University Press, 2019).
[11] Varriano, *Wine: A Cultural History*, 63–6.
[12] Varriano, 53.
[13] Phillips, *9000 Years of Wine*, 49.

[14] Ibid.
[15] Tim Unwin, *Wine and the Vine: An Historical Geography of Viticulture and the Wine Trade* (London: Routledge, 1991).
[16] Claude Chapuis and Steve Charters, "The World of Wine", in *Wine Business Management*, eds. Steve Charters and Jérôme Gallo (Paris: Pearson France, 2014), 13–23.
[17] Jean-François Bazin, *Histoire du vin le Bourgogne* (Paris: Ēditions Jean-Paul Gisserot, 2013), 21–2.
[18] Rod Phillips, *French Wine: A History* (Berkeley: University of California Press, 2016), 53–4.
[19] S. Ouvrard and I. M. Taplin, "Trading in Fine Wine: Institutionalized Efficiency in the Place de Bordeaux System," *Global Business and Organizational Excellence* 37, no. 5 (2018): 15.
[20] The early English preference was for a rose-colored wine that came to be known as claret (*clairet* in French). It was probably the wine that best could withstand the long journey involved in shipping to England at that time.
[21] Phillips, *French Wine*, 86–7.
[22] Dewey Markham, *A History of the Bordeaux Classification System* (New York: Wiley, 1998).
[23] Ouvrard and Taplin, "Trading in Fine Wine," 16.
[24] See Pierre Marie Chauvin, "Globalization and Reputation Dynamics: The Case of Bordeaux Wines," in *The Globalization of Wine*, ed. David Inglis and Anna Mari Almila (London, Bloomsbury, 2020), 104–5.
[25] Ouvrard and Taplin, "Trading in Fine Wine," 16.
[26] This notion was discussed extensively with château owners on both the left and right banks when my colleague Stephane Ouvrard and I conducted interviews over the past four years. We were particularly interested to see what power in price setting the respective owners had when it came to selling their wine, and not surprisingly the higher their ranking, the greater the pricing power.
[27] Phillips, *French Wine*, 153.
[28] James Simpson, *Creating Wine: The Emergence of a World Industry, 1840–1914* (Princeton: Princeton University Press, 2011).
[29] Jean-Michel Chevet, Eva Fernandez, Eric Giraud-Héraud, and Vincente Pinilla, "France," in *Wine Globalization: A New Comparative History*, eds. Kym Anderson and Vicente Pinilla (Cambridge: Cambridge University Press, 2018), 67.
[30] Simpson, *Creating Wine*.
[31] Ibid., 58.
[32] Chevet et al., "France," 87–8.

Chapter Two

Missionaries and Adventurers: The Early Years of California Wine

We have seen how religion played an important role in the development of wine from ancient times onward. As noted in the last chapter, much credit is owed to the Cistercian and Benedictine monks in Burgundy who put winemaking on a firm footing in the Middle Ages in France. When we look at wine's origin in California, religious groups are again important, but this time Franciscan missionaries were the pioneers. When these missionaries moved up the west coast of what is now California, then Spanish Mexico, in the late eighteenth century, their ostensible purpose was to transform the spiritual life of the native population. Imbued with this sacred fortitude but also cognizant of the need for survival, they built missions that would serve as the focus for their religious activities. In the area surrounding the missions they introduced systematized agricultural production as part of their self-sufficiency goals. These missions dotted the coastal area from Mission San Diego de Alcala in the south to the northernmost Mission San Francisco Solano in Sonoma, founded by Father Jose Altamira in 1823. In each mission, crops were planted using labor from the indigenous population, and attempts were made to convince such people of the merits of a quasi-"civilized" religious existence. When harvests were good, which they frequently were in the coastal California region, such benefits were presumed to be self-evident. Moreover, regular and consistent food production from the mission farm could improve the physical health of the local population. Since the missionaries had an overriding spiritual mandate, and given the centrality of wine for their sacred rituals, it is not surprising that the Franciscans also planted vineyards, probably using vines that had their origins in sixteenth-century Europe.

The first documented evidence of vines being planted by the Franciscans was at San Juan Capistrano Mission in the south. This was in 1779, with the first vintage produced in 1782.[1] Other missions going north up the coast probably had similar plantings, but evidence is scarce. The grapes that were planted have come to be known as "Mission grapes." They were of the vinifera varietal, were large, and yielded vigorous crops. Furthermore, they were apparently good to eat, and made nice raisins when dried. One can only surmise what they tasted like since evidence is anecdotal, but as wine historian Charles Sullivan aptly notes on the last surviving examples of this variety, "the wines made from them intended to be red are usually deficient in color and dull to the taste. Dry white wines are equally dull and harsh and tend to oxidize and darken easily."[2] Hardly tasting notes that inspire a contemporary oenophile to sample such a product! But, as we have seen in previous chapters, tastes change and are context-driven.

We know little of the amounts that were produced in the various missions, although comments by travelers suggest that there was enough for the secular table as well as the altar. Those same travelers commented on the quantity and quality ("plenty of good wine during supper," according to one visitor to San Gabriel Mission in 1826[3]), which, given Sullivan's earlier comments, perhaps suggests a less-discriminating clientele who, after an arduous journey, were more than happy to enjoy any alcoholic beverage, or that San Gabriel actually had a good winemaker monk. Other than these anecdotal comments, little is known of the commercial extent of the missions' wines. Some trade probably occurred after Mexico's final independence from Spain in 1821, and the missions conceivably sought some sales as a way of further subsidizing their religious activities. However, this came to an end in 1833 when their properties were secularized by the state and the land distributed among the indigenous population as well as new settlers attracted by the prospect of available land. The monks' activities were now limited to spiritual care as they became de facto parish priests.

The person responsible for distributing the land of the Sonoma Mission in northern California was Mariano Vallejo, a lieutenant in the San Francisco Presidio. He was ordered north of the San Francisco Bay in 1834 and charged with breaking up the huge clerical properties in that surrounding area (including parts of Napa), returning the land to the local Indian population. Himself the beneficiary for his actions of the huge Petaluma Rancho, which was intended for distribution but retained as his personal property, he did hand out much land in the Sonoma area, but mostly it went to his friends and those who had assisted him in developing the area as a site of large ranches. Alas, the local Indian population proved far less fortunate beneficiaries of his actions and largesse.

48   Chapter Two

Fig. 3.1. California Mission

Source: author's sketch

One of his acquaintances was North Carolinian and venerable mountain man George Yount. In return for helping Vallejo make Sonoma into a social and political administrative center and providing valuable building advice, he was granted the former mission's Caymus Rancho in the area that is now central Napa Valley, but was then largely unoccupied.[4] He moved to the twelve-thousand-acre area in 1838, growing numerous crops (including an orchard) and planting vines which he obtained from Vallejo's estate. The vineyard he planted is near to what is now Yountville (the town named after him in the central part of the valley), and represents the beginning of systematic wine production in Napa.

Smaller tracts in the southern part of Sonoma and Napa were granted during this time, but the next big transfer was given to Edward Turner Bale, a naturalized English surgeon who gained his property in 1841.[5] Adjacent to but bigger than Yount's property, it extended northward to beyond what is now Calistoga. Even though he planted vines, he also sold off parcels to others who were interested in the potential of viticulture. By making land available to newcomers, he not only profited from such sales but also set in place the development of the northern half of the valley into an area where people were experimenting with crops, including grapes on smaller plots of land. This pattern of small plots alongside large would continue for ensuing decades.

In 1848 Mexico ceded much of its northern territory, including California (hitherto Alta California) to the United States. It was largely an area with small towns surrounding missions and was ripe for development, if there were sufficient attractive prospects for newcomers. Those that came did so to take advantage of the land giveaways, but it was the discovery of gold in 1849 that unleashed a migratory inrush with more than sixty thousand arriving from around the world in search of their fortune. The consequences of this influx were twofold. First, demand for alcoholic beverages (including wine) inevitably grew with the increased male population who inhabited the taverns and saloons in the mining camps at the end of the day, and were not noted for their purchasing frugality. Second, when the gold began to run out, many sought opportunities in the emerging vineyard areas to the northwest, sometimes bringing the money they made from prospecting and investing in land. Others who were less fortunate came seeking opportunities as laborers. Even though agricultural opportunities abounded, it was recognized that wine could easily be grown in most parts of the state, so some were attracted to viticulture. Since many immigrants were of German and French background, wine was at least a familiar entity that they could try their hands at, so perhaps not surprisingly many of them were early wine pioneers.

By the late 1840s there were vineyards throughout much of coastal California, although the majority were in the southern part of the state.[6] However, after 1849 the market for wine was in the north, so production shifted gradually to that area, eventually displacing the south as the primary wine region in the state. But before that conclusively happened, a vigorous trade in the shipment of wine from the Los Angeles area (where the wine was made) to San Francisco (where it was sold and drank) had emerged. This incipient wine market created commercial opportunities for enterprising traders who purchased wine in the south then shipped it north for sale. One of the most notable and profitable trading houses was the firm of Kohler & Frohling, formed by two German musicians who raised money in 1854 and bought a small vineyard of mission vines in Los Angeles. From this early venture they went on to dominate the trade of wine from southern to northern California, ultimately consolidating production in the south by buying wine from numerous growers, then shipping it north for storage and eventual sale from their warehouses. The commercial scale of such activities was impressive, and they remained a dominant force in the industry, eventually shipping wine to the East Coast. Over seventy thousand dollars of wine was sent to New York and Boston in the 1860s, and they continued to gain accolades for its quality.[7]

While many vineyard owners persisted with Mission grapes, others were experimenting with additional vinifera varietals. One such pioneer was J. W. Osborne, who in 1851 purchased a large tract just south of Yountville in what is now Oak Knoll. A former sea captain and avid horticulturalist, he planted a variety of vines to determine which were best suited to the climate and soils of the valley. Many of these varietals are not well known today,[8] but one, Zinfandel (then called Zinfindal), has remained important. Osborne was at the forefront of a nascent movement to replace the high-yielding Mission grapes with varietals that would produce quality wines, albeit with lower yields. This did not always make him a popular advocate for the industry, as many vineyards were intent on maximizing output to stay profitable. From his vines, however, he managed to produce wines that were acknowledged to be of high quality, winning awards at northern California fairs and thus demonstrating the merits of his approach. Along with another wine pioneer, the Hungarian Agostin Haraszthy, he cofounded the Sonoma-Napa Horticultural Society in 1859. This organization attempted to systematically address various techniques for viticulture and viniculture, and share such formal knowledge to interested parties. Several years earlier, in 1854, the California State Agricultural Society had been chartered with a mandate to improve the dissemination of grape-growing knowledge through its publication *Transactions*. In fact, it commissioned Haraszthy to write a "Report on Grapes and Wines of California" in 1858.[9] In this document, he provided extensive practical notes on growing, tending vines, and winemaking, and it was widely distributed throughout the state. Elsewhere, the San Francisco-based weekly magazine *The California Farmer* continued to promote winemaking, but all too often relied on information applicable to midwestern and eastern vineyards. This was unfortunate given the significant climatic differences between the regions, which rendered the relevance of such edicts questionable. Haraszthy's report was however more pertinent to local conditions, and gained traction among other newcomers eager to avoid the pitfalls of earlier pioneers.

One of the first indications of institutional support for the wine industry came in 1859 when the state legislature introduced a four-year tax exemption for all new-planted vineyards. This was formal recognition of the time lag between planting and production before an established revenue stream occurred for new wineries.[10] The tax law was followed in 1861 when the legislature introduced the "Commission upon the Ways and Means best adapted to promote the Improvement and Growth of the Grape-vine in California."[11] As the title suggests, it was designed to focus on procedures that could be beneficial to wineries (and olive growers) by aggregating the most pertinent information available at that time. Together with the earlier fiscal policy, this demonstrated the fledgling state's willingness to provide support for an emerging wine industry, helping to further the relocation of former miners who were intent on settling in the area as well as recognizing an agricultural sector deemed crucial for the state's future development. This evolving institutional framework, albeit limited in its inception, would nonetheless lay the foundations for other attempts in subsequent years to create a formal body of knowledge for industry practitioners.

Napa's Quality Emerges

By the 1860s, the northern part of California was proving to be a popular place to grow grapes and make wine, with numerous microclimates that made it more favorable than the southern part of the state. Sonoma was the principal area where most vineyard planting occurred, whereas adjacent Napa saw a mix of orchards, nut crops, and vines. Even with limited institutional support, most winemaking was hit and miss, and practitioners had rudimentary skills at best. It remained a work in progress. In retrospect, we can recognize the beneficial aspects of the judicious combination of soil and climate in the area, but at the time such knowledge was incomplete and evolving. People did not really know what grape varietals grew best in what soils, and the actual trade of winemaking was still very much in its infancy. Not surprisingly, many persisted with planting mission vines, since that was the known grape, and a market for it, often as a blend, persisted.

As with any new venture resources, both financial and technical knowledge can be crucial for success. Winegrowing was no different since the upfront costs are high, and returns on the initial investment take three to four years before they are realized. The newcomers to California's developing wine sector came from different backgrounds, and there was considerable variation in the resources they possessed. Many were Europeans who, like their counterparts elsewhere in nineteenth-century United States, were in search of ill-defined opportunities. The attraction of northern California was the availability of inexpensive land, although the journey there probably dissuaded all but the most enthusiastic. Some might have known something about grape growing or agriculture in general, while others were adventurers looking for a new start and the land was the draw. Vineyards seemed an interesting prospect for both groups. From available records it appears that, of those that did set up in the emerging commercial wine industry, success often came when individual resources cushioned their early endeavors. Wealth did not make you immune from failure, but it gave you more leeway to experiment. Furthermore, it meant you could try new grape varietals to determine what was best suited to the area, and not be wedded to extant practices.

One notable group of newcomers with such resources were German immigrants Jacob Gundlach, who went to Sonoma, and Charles Krug and Jacob Schram, who went to Napa. With Hungarian Haraszthy, they were willing to experiment with imported vines and finally make the break from Mission grapes. In 1856, Haraszthy bought and developed a new 560-acre property northeast of the town and mission of Sonoma and named it Buena Vista winery. He imported fourteen thousand vines that consisted of 165 varieties, grew more than twelve thousand vines in his nursery, and planted vineyards for Gundlach and Krug as they developed their new properties.[12] The other Napa pioneer, George Crane, planted Mission grapes in his new three-hundred-acre vineyard just south of St Helena in 1859. When half of his crop died several years later he replanted with foreign cuttings of varietals such as Riesling and Sylvaner, which when harvested in 1862 were, according to the leading Sonoma newspaper, "as good as anything made on their side of the mountain."[13] Adding to the vineyard growth in Napa was Sam Brannan, who bought a large tract of land near Calistoga in 1862 and planted twenty thousand cuttings he had brought back from a trip to France in 1860. In 1863, Schram experimented with as many varieties as he could afford on his hillside site, eventually proving that good wine could be made outside of the valley floor.

By the end of the 1860s, several successful vintages had demonstrated the quality and relative consistency of wines produced in Napa. Climate and soil appeared to be favorable, and more plantings followed. Most were of foreign varietals, as Mission grapes were finally recognized as unsuitable for making wine with the best taste. New plantings continued in Napa throughout the 1870s, with varying harvest results and wine quality. Between 1869 and 1871, Krug's production went from twenty to fifty thousand gallons, and another German immigrant Jacob Schram planted more vines to meet what he perceived to be a growing demand.[14] Record vintages in 1872 were followed by poor ones in 1874, a further indication that this was an agricultural crop subject to the vicissitudes of the weather, no matter how perfect the soil and site conditions were. Zinfandel was hailed as the high-quality crop, but a large variety of other grapes, such as Malvoisie, Angelica, Muscat, Burger, and Charbono, were grown, some suitable for rosé style wines while others were naturally sweeter. But even with a good harvest, demand for wine was poor in some years, especially in the recession years of the late 1870s. Wine was still a commodity with all of the uncertainty of supply-and-demand disequilibrium in fluctuating market conditions, and a distinctive regional wine culture that would sustain the market had yet to fully emerge.

While the individuals mentioned above have gone down in history as the wine pioneers, many vineyards were small, with the owner as the winemaker. Even by the 1890s, the average size of a grower's property in Napa and Sonoma was sixteen acres. Of the 284 listed wineries in 1891, forty-one were in excess of 125 acres, whereas 170 had less than twenty-one acres.[15] In the smaller properties, tonnage was modest and manageable for a family venture. Using rudimentary equipment to crush and ferment the grapes, the vinified product was most likely stored in casks (generally California redwood, which was in plentiful local supply) and then sold to a merchant in San Francisco.

Viticulture knowledge varied tremendously among the small wineries – some saw themselves as simple farmers, others as more dignified horticulturalists. In either case, they were often experimenting by trying to discern appropriate techniques, and doing it with imperfect information and fractured viticulture knowledge. And then they had to sell the finished product in a rapidly fluctuating marketplace in which San Francisco wine merchants were becoming more and more influential. They lacked production-scale economies and their small size limited their marketing ability. It is not surprising that many clung to varietals and production techniques that provided the largest yield. Notwithstanding poor-harvest years which hurt farmers, even years with high production could be problematic financially. A bumper harvest typically resulted in excess capacity, oversupply, and depressed prices in a still-evolving market where consumer tastes remained uncertain. Even though some wineries pursued quality, others simply viewed production through the typical agricultural lens of productivity and sought to maximize crop quantity. Such actions often led to deteriorating quality, which adversely affected Napa's reputation. Continued comments on the poor quality of Napa wine during this period, alongside others that were more complimentary, indicate the contrasting reputations that persisted.[16]

The quality versus quantity debate would continue for the next few decades. That Napa could produce good-quality wines was increasingly acknowledged during the 1880s. The San Francisco newspaper the *Daily Alta California* noted in 1880 that, "Napa is now the leading wine-growing county of California ... and St Helena has become the centre of the most prosperous wine district in the state."[17] Three years later, the *Sonoma Index* commented on how merchants were paying higher prices for Napa wines than those from Sonoma, suggesting that the public perceived the quality of the former to be superior.[18] But there was also much wine that was faulty, partly because some winegrowers still blended with Mission grapes, and simply did not understand that overcropping could dilute the finished product, or picked unripe grapes in order to get to market early. In other words, many continued to face the harsh financial imperative of production that would guarantee a revenue stream and thus sought to maximize yields. Since the market (in this case the merchants) condoned such behavior by not being excessively discriminatory over quality, there was little incentive for such growers to change their behavior.

As for the merchants who were responsible for selling much of the wine, they had consolidated their position in the market. Building on the early commercial success of traders such as Kohler and Frohling, merchants had become a powerful group, well-resourced and in a commanding position to dictate prices since they were the lynchpin in the distribution of wine. As the volume of wine increased, from around 2.5 million gallons in 1870 to ten million in 1880 and eventually almost twenty-four million by the early 1890s, their intermediary position was crucial for the continued development of the industry. They needed a regular supply of wine, much of which was then blended, so quality was less of an imperative for them. They did use their power to depress prices paid to growers when there was an oversupply of wine, something that larger growers noted which subsequently influenced their decisions to consider focusing on quality rather than yield.

Access to collective knowledge about grape growing and winemaking that could be beneficial to industry development was random at best during this period. Informal groups of winegrowers met from time to time to discuss salient issues, but there was never a formal framework that could more systematically mitigate the problems. Occasionally, there were attempts to address local technical problems, and such activities alerted others to the need for something more formal at the state level. The St Helena Winegrowers Association was one such group, and in one notable instance of collective behavior they pledged to not sell grapes to producers who used added sugar (a process known as *chaptalization*) in production in order to increase alcohol levels.[19] Such actions appeared to be somewhat beneficial, but the long-term consequences were unknown. Members of the association sought to discourage it in order to obviate potential problems but also to safeguard the emerging reputation of Napa. At issue here was an early recognition that the overall reputation of the district was more important than individual self-interest. While some growers balked, others (like Charles Krug) forcefully argued that measures such as this ban would be important for the industry to continue its claims for oenological legitimacy.[20] Such a leadership role by key individuals would reoccur in subsequent decades and be important in spearheading a more formal governance structure. Notably, Charles Krug continued to play an important role in this association as it evolved as an informal forum for winegrowers in the area.

At the state level, ad hoc organizations had a wider umbrella, often emerging to protest tax and tariff issues. In 1862, the California Wine-Growers Association was formed, ostensibly to encourage industry growth, but most ardently to discourage federal taxes on wine.[21] It lasted barely a year but was resuscitated in 1866 when the tax threat re-emerged. In 1872, the California State Viticultural Society was created, and while frequently preoccupied with taxes and tariffs, it did preserve in its mandate committees to improve wine cultivation as well as hold annual fairs in which wines could be presented for comparison and prizes. It highlighted quality, but its role as an advocacy agency seemed more beneficial to wine merchants than winegrowers, and it lacked legitimacy in the eyes of small farmers. The latter were frequently denied input, and viewed many of the proposed viticultural improvements as beyond the scope of their capabilities.

The respected and powerful growers, particularly in Napa and Sonoma, nonetheless pressed for more state resources that would enable them to market their wine nationally. Claiming to be able to produce quality wines on a fairly consistent basis, they were leveraging the growing reputational benefits by articulating a variant of the French terroir argument. By highlighting the economic possibilities of wine, they were appealing to state politicians who might be eager to capitalize on the commercial viability of this industry. This pressure created the Board of State Viticultural Commissioners in 1880. As part of the California State Agricultural Society, the board was given a wide mandate to promote viticulture in the state, and was made up of members from each of the state districts. Prominent winery owners such as Krug served on the board, giving an effective voice to the wealthier segments. How instrumental it was in its promotional endeavor is difficult to determine, although it did provide the focal point for discussions about the emerging phylloxera epidemic sweeping the vineyards in the late 1870s.

In 1875 the University of California appointed Eugene Hilgard as professor of agriculture, and in that capacity he made frequent comments about the nascent wine industry in the North Bay in his remarks to the State Viticulture Society. He was a proponent of labeling California wines to note their place of origin in what he envisaged was a growing global marketplace for wine, thus presaging subsequent regional geographic origin demarcations. At the university he built a model wine cellar with money from the state, and proceeded to conduct extensive experiments on appropriate varietals and general winemaking. His embrace of chemistry and scientific objectivity in his systematic investigations shocked many, but his persistence proved of lasting benefit to the industry since he was able to demonstrate the effectiveness of numerous techniques, especially in the fermentation process.[22]

His advocacy of the science in winemaking was salutary, but he was also forthright when it came to quality issues and the need to learn from the mistakes of the past. In an address to local winegrowers in St Helena in 1880 he was adamant in his condemnation of those who continued to pursue quantity because he believed this was counter to the true potential of the area's wines. He opposed blending with Mission grapes, and claimed that grape quality was contingent on soil, thereby encouraging growers to send him soil samples for tasting. This far exceeded earlier attempts to distinguish property attributes based simply on geography. As a non-winegrower, his endorsement of Napa and its potential was powerful. As Charles Sullivan notes when reporting Hilgard's final comments to the group, "Now is your golden opportunity; and if you act wisely, energetically and unitedly, you are sure of success."[23] A university specialist was now adding the veneer of technocratic legitimacy to what many winegrowers had been claiming for years.

In the years following his appointment at the university, Hilgard coordinated the writing and dissemination of numerous technical reports that would be appropriate to the climate of Napa and Sonoma. In awe of European winemakers and their techniques, many wine growers in northern California replicated European practices by letting their grapes hang on the vine as long as possible in the autumn prior to harvest. This technique was used in Europe where cool autumnal days are useful for ripening the grapes, but in California the weather remains hot and delaying harvests is unnecessary. In fact, harvesting grapes too late, as many growers were doing, resulted in high alcohol and sugar levels along with low acids. While twenty-first century winegrowers have managed to mitigate these problems and in many cases celebrate the intensity and stylistic benefits of such a process, back in the 1870s the wines made under such conditions were often too harsh and earthy flavored. What worked well and was standard practice in France was not always relevant or appropriate for northern California. Hilgard's reports helped confirm this. The veracity of these technical reports managed to synthesize Old World knowledge and the peculiarities (and notable assets) of the northern California climate.[24] Hilgard also incessantly preached the virtue of cleanliness in wineries, forcefully arguing that systematic and vigorous cleaning was crucial to eliminate harmful bacteria that could spoil wines.

Another proposed technical innovation focused on fermentation techniques proved valuable to winegrowers. High temperatures during fermentation (and subsequent storage) frequently impaired the quality of the wine. To overcome this problem, wineries were encouraged to dig into hillsides and introduce rudimentary gravity-flow techniques (grapes were crushed and fermented on the top floor, then allowed to flow to a lower, cooler floor for finishing). This mitigated the problem and established practices (and basic construction details) that remain in many wineries to this day. One of the wineries that used these new techniques was Eshcol (renamed Trefethen after that family purchased it in 1968), founded in 1886 by a Scottish sea captain Hamden McIntyre just north of the town of Napa. He subsequently helped design other gravity-flow-system wineries in the area, including Far Niente and Inglenook. The technical revolution of the 1880s would prove of lasting significance for Napa since it enabled winegrowers to focus more on quality and better understand how that could be consistently achieved.

While Mission grapes were increasingly removed from production, no signature varietal had consistently emerged to replace it. Zinfandel was proving popular, as was White Riesling, Grenache, and Mourvedre. New plantings of these continued to increase into the 1890s as more varietal performance information became available. At the forefront in the promulgation of this information were key agricultural commissioners. One of the most influential members of the Board of State Viticultural Commissioners was Charles Wetmore.[25] Not only was he responsible for tirelessly collecting data and publishing reports and instructions during the 1880s, he also recognized the great potential that Napa had for making high-quality wine, and suggested that growers consider planting Bordeaux varietals such as Cabernet Sauvignon. Until then, it was not a grape that had been much considered.

Official affirmation of Napa's quality came at the 1888 State Viticultural Convention in San Francisco in which red and white wines from throughout the state were carefully selected and evaluated by judges, all of whom were professional wine experts. The final results firmly placed Napa producers at the top, receiving a total of twenty-one first awards. Judges commended the producers of Bordeaux varietals, and Napa had the top Cabernet Sauvignon. The local press extensively publicized the results, and the competition established Napa's early reputation for producing premium table wines.[26]

58   Chapter Two

Fig. 3.2. Eshcol Winery ca. 1890s

Source: author's sketch

Improved quality continued to benefit from the presence of well-resourced newcomers who entered the industry with a commitment to excellence and the financial means to implement appropriate new techniques that might get them there. One such person was former sea captain Gustav Niebaum, who had amassed a large fortune and was intent on buying land for a winery. He acquired Inglenook winery near Rutherford in 1880. The winery had been established by William Watson in 1871, but Niebaum pulled out the mission vines, planting Bordeaux varietals in addition to white Riesling, Pinot Noir, and Zinfandel, hired capable experienced managers to oversee production, and extensively rebuilt the winery. When it opened in 1891 it was a state-of-the-art facility in which his best wines were estate bottled for direct sale to consumers, some of whom were on the East Coast. This basic branding strategy (he developed a distinctive diamond trademark label for his bottles) was part of a wider campaign to introduce his wines at public events and ensure that the press received as much information as possible about his activities. As wine historian Charles Sullivan aptly notes, "Niebaum's great service to the California wine industry was his ability to present a dependable line of wines at the premium end of the price spectrum, and to do it with just the right balance of showmanship to appeal to eastern customers."[86]

Overproduction, Economic Downturn, and Merchant Consolidation

Improved quality and consistency in Napa wine was commendable, but selling that wine could at times prove difficult. The area's reputation was growing, but the market for premium wine was the East Coast where Old World wines still had the upper hand, despite declining supplies following the European outbreak of phylloxera. California wines were seen as somewhat novel, possibly interesting flavor-wise, and occasionally cheap but also frequently bland or earthy. They were most likely sold as blended generic wines (e.g. Red or White Burgundy, Chablis), with only occasional attempts at brand differentiation or geographic origin specification (i.e. Napa).

It had been thought that the nation's growing southern European immigrant population would take naturally to California wines, but such groups were often poor and frequently sought the cheapest wine available. Similarly, the burgeoning male population of nearby San Francisco certainly had alcohol devotees, but they too gravitated to the lowest common denominator when it came to wine. The reality was that the USA was still not a nation of wine drinkers, with many of the aforementioned males most interested in seeking out cheap hard liquor as the beverage of choice. Furthermore, hotels and restaurants were unenthusiastic purveyors of wine, and in such establishments no attempt was made to differentiate by brand or region (wine for sale was frequently listed as either white or red).

The predicament for Napa producers, as well as those in nearby Sonoma and other adjacent areas in the north, was that, despite being able to make premium wine, it was difficult to consistently sell it at a premium price. This was further compounded when productive harvests in the last two decades of the nineteenth century coincided with economic downturns. Variability in sales had always been somewhat cyclical, but the long depression in the 1890s proved dire for many wineries. The resulting declining market for wine in general, together with its oversupply, inevitably led to downward pressure on prices. All too often, the prices paid for grapes did not cover the costs of picking, let alone other overheads. This even affected some of the biggest producers, such as Charles Krug who went bankrupt in 1885, burdened by debts associated with overexpansion.

Solutions to market volatility typically involve actions by key actors to manage supply-and-demand disequilibrium by imposing restrictive practices and increasing concentration. In the case of California's wine industry, efforts to cooperate and coordinate activities to enhance pricing power, when they did occur more formally, came from the wine merchants. Competition was fine when prices and demand were high, but both declined significantly during the 1890s. An attempt to rectify this problem came in 1894 when seven of the state's most powerful wine merchants, all based in San Francisco, formed the California Wine Association (CWA).[28] Many merchants were also large winegrowers, following a pattern several decades earlier when Kohler and Froling had bought vineyards in the Los Angeles area, and who were now founding members of the new association. Also, some of the larger winegrowers throughout the state had created their own distribution network to get their wine to East Coast markets. To further such practices, they joined the CWA since it provided the umbrella coordination that their individual efforts lacked. The best example in Napa was Greystone Cellars and, in Sonoma, Glen Ellen. This backward-and-forward integration across the key facets of the supply-and-distribution chain was seen as a way to leverage scale efficiencies as well as coordinate pricing strategies and manage supply. Through their size, they were able to exercise control over most aspects of wine sales. But power within the CWA remained firmly in the hands of the merchants, and their interests took primacy over those of growers.[29]

Not surprisingly, the apparent monopolistic tendencies of the CWA irked growers, who were not part of the organization since they remained price takers, typically bereft of any pricing power that would ameliorate their cost concerns. Anticipating the adverse consequences of this asymmetrical relationship, they too sought refuge in an associational network, and in the same year formed the California Wine Makers' Corporation (CWC). Designed to heighten their own bargaining power with the CWA, the growers were also looking for ways to realize the financial benefits of producing better-quality wine. When wine was sold to merchants it was typically blended and then shipped to the East Coast, with a standard price paid to the growers, often regardless of its quality. This frustrated some winegrowers, especially in Napa, who knew they could make good wine. To rectify this problem, the corporation differentiated according to quality and sought appropriate (higher) prices for such wine. This meant bypassing the CWA by offering an alternative route to market for growers. In doing this they directly competed with the CWA. But the CWA proved a formidable organization, and its response to this challenge was to lower selling prices, significantly undercutting those charged by the growers in their alternative channels. Even though many growers believed their wine was of sufficient quality to merit a higher price, the general public had yet to come on board with that sentiment. With sales stymied, the growers' association collapsed in 1899. Ironically, though, some of the large growers in the CWA had begun to recognize the branding potential of separating wines on the basis of quality that could provide a price premium. Since wine was shipped in bulk (barrels) from the winery to the distributors, it was often difficult to retain a sense of individual identity for the winegrower. Yet, it was logistically difficult for them to find a way to efficiently provide a parallel distribution channel for their specific wine, even if they could persuade other CWA members of the merits of this approach. The more powerful the CWA became as the conduit for the shipment and sale of wine to consumers, the more difficult it became for winegrowers to break ranks and develop an alternative distribution model.

As the CWA increased in size with new members joining, it further consolidated its position in the wine marketplace, thus shaping the evolution of the industry during the first decades of the twentieth century. It became an effective marketing voice for California wine, but often at the expense of individual wine growers' identity since the latter's wine was subsumed under the generic category of California wine (sometimes simply listed as Burgundy). By coordinating the interests of the powerful actors in the industry (wine merchants), it did provide governance structure and some market stability. Production (and demand) gradually increased in the first decade of the new century from twenty-three million gallons in 1900 to forty-five million by 1910. Quality was improving, partly due to the continuous spread of technical information on winemaking from the university and a better understanding of what varietals were best suited to the California climate. But many winegrowers still faced uncertainty over their subordinate position in the sale-and-distribution nexus and its impact on their profitability. To overcome this, they needed a more-effective voice and the means to benefit from the value-added potential of their wine. Relying on a collective reputation for California wines in general did little to differentiate their distinctiveness. What was necessary was a more assertive individual focus on individual brands, but this proved difficult under the constraints of the CWA.

Napa Resurgent

Excellent vintages in 1899 and 1900 showcased the quality of Napa wine (as well as those from Sonoma and the south bay area around Los Gatos). Notwithstanding the tribulations and acrimony of the fight between the CWA and the CWC, new plantings in Napa were replacing land hitherto devoted to fruit orchards.[30] New industry investors seemingly took advantage of technical improvements and had the benefit of more widespread information on grape varietals when making planting decisions. While Zinfandel was frequently derided for its lightness, it was easy to grow and had good yields. But it could not command the anticipated price premium that fuller, richer wines such as Cabernet Sauvignon could attain. Some growers as well as technical experts saw Cabernet Sauvignon as a varietal that could flourish in the northern California climate and produce a quality wine, probably similar to that found on the Left Bank of Bordeaux. But as long as varietals were blended before sale, this was often a moot point. And the continued power of the CWA proved a further disincentive for excessive experimentation since it remained equivocal about quality. When change did eventually come, it was the result of a pest-born disease that destroyed many existing vineyards, forcing new plantings.

The phylloxera outbreak in the 1880s and early 1890s devastated many vineyards in northern California, and vines were subsequently ripped up. But the replanting with disease-resistant rootstocks (another benefit of the increased technical information and scientific method) enabled winegrowers to try different varietals. Some focused on Cabernet Sauvignon as well as whites such as Sauvignon Blanc and Sémillon. Others simply replanted disease-resistant rootstocks of the varietals they had before. While by no means a revolutionary change, the new varietal plantings were nonetheless a testament to the evolving understanding of the soil and climatic nuances of the valley and its suitability for certain grapes. Cabernet Sauvignon appeared to have good quality/yield ratios that rewarded those who experimented with this varietal planting.

Napa newcomer George de Latour (who trained as a chemist in France) acquired land in 1900 and imported rootstocks from France, planting a wide range of varietals, mostly those previously grown in the area. But within a decade he recognized the true potential of Cabernet Sauvignon and devoted much of his newly renamed vineyards, BV #1 and BV #2, to that varietal. Furthermore, in 1903 the USDA established an experimental site at To Kalon vineyard near Oakville where resistant rootstocks were planted.[31] Again, Cabernet Sauvignon was proving to be one of the most notable of the varietals planted in terms of quality/yield ratios. Such a combination proved seminal with the 1904 vintage, and more remarkable vintages followed, especially in 1908, but again volume was down. Despite overall poor quality because of overcropping, some winegrowers nevertheless focused on maximizing quality from their vineyards, albeit with the inevitable low yields. They produced what became remembered as stellar vintages, and Napa reds received much of the publicity.[32]

Independent producers were often at the forefront of these successful vintages, resisting the pressure to go through the CWA. Capitalizing on the emerging reputation of Napa, they developed alternative direct distribution channels to sell to discerning customers, often on the East Coast. But they too were sometimes caught in the yield/price wars. New plantings and the growing maturity of vines planted after the great 1900 harvest often added to overall production, and the subsequent oversupply. Since lower tonnage per acre (generally less than two tons) generally translated into lower income for growers, even if they could argue that they were quality grapes, they still suffered price depressions. Not surprisingly, many were unwilling to eschew old methods and opted for higher yields. But higher production in turn further depressed prices unless demand for wine increased, and that was not always the case during these years. It was a vicious cycle with no easy escape for many growers.

Despite its growing reputation, Napa continued to produce many poor-quality wines that met the minimal requirements of blending into generic categories of bulk wine. This was the quintessential commodity product. By the turn of the century, the CWA had started adding brandy to much blended wine that was deemed unstable.[33] This fortified wine kept longer and its sweetness appealed to a certain market segment, but such actions effectively postponed many technical improvements in winemaking since flaws could be more easily disguised through this distillation. This favored bulk producers from the hot areas of the central valley who struggled to make a quality wine more than their Napa counterparts. It also meant that there was often little incentive for the average winegrower to break from standard practices because the upfront costs for replanting or even reducing yields to gain a better quality were high. However frustrated they might have been with the vicissitudes of the market, the *status quo ante* remained prevalent when changes were aired. They continued to lack pricing power and remained subordinate players in market channels. Prices might be higher if demand was great, but for this to occur the quality had to be guaranteed and recognized. Such a guarantee would often be difficult for a small grower, and they certainly lacked the marketing prowess to broadly advance their reputation.

This somewhat bifurcated market characterized Napa in the first decade of the twentieth century. Some large growers were recognized for the quality of their wines, and while collective reputational benefits might accrue to others, many producers maximized yields and relied on the CWA to be a consistent buyer for their product, regardless of its quality. The CWA meanwhile continued to market California wine as a value proposition, although it did occasionally note the distinctive quality of certain producers.

The final pre-Prohibition event that put Napa in the spotlight was the 1915 Panama-Pacific International Exhibition (PPIE). As with many of the World Fairs of the nineteenth century, this event was designed to showcase the many innovations of early twentieth-century America as well as demonstrate how, phoenix-like (literally), San Francisco had emerged from the ashes of the earthquake and fire that ravaged the city in 1906. Given the importance of agriculture for California's economy, wine not surprisingly took a prominent place in the proceedings. Images of cellars and wineries were presented alongside wines to create a lasting impression of the almost Arcadian splendor of rural life in the state. An international jury was assembled and awarded many of the prizes to Napa wines, including twenty gold medals for Inglenook. How good the wines were is difficult to discern. Conceivably, the growing reputation of Napa vis-à-vis other regions in California conferred a certain tasting bias toward wine from that area. In other words, one wonders whether the judges were acknowledging the growing brand recognition of Napa, the product of consistent marketing efforts by the CWA, rather than objectively judging the wines. There is no certainty as to whether tastings were blind, and one cannot help but recall how the 1855 Bordeaux classification merely formalized the extant recognition of quality wines based on price and early reputation. Since more people now acknowledged Napa wines to be of high quality, perhaps the judges were willing to formally endorse that. Ultimately, we will never know, but the formal acknowledgment of Napa's quality conferred by this tasting was a milestone in the valley's evolving reputation.

If Napa did make quality wines, which was now recognized, why weren't more growers embracing the operational philosophy underlying it? Independent producers might be able to ride the coattails of the CWA juggernaut, but their efforts to create an alternative albeit informal cooperative structure were dimming, and many still lacked the bargaining power to get their wine to market on favorable terms. Technical innovation came via formal channels and official reports, and quality was a virtue consistently preached in such documents. But unless the CWA was willing to renounce the practices (such as blending and adding brandy) that had made it wildly successful, and start rewarding low-volume, high-quality producers with prices that encouraged them to eschew quantity production, not much would change. Those growers such as de Latour and Inglenook who had developed their brand based on quality would help build the collective reputation, but only insofar as others would be able to embrace a similar operational philosophy. As we have seen, this often proved a difficult proposition.

Prohibition and the Collapse of the Industry

For much of the nineteenth century, the United States had a love/hate relationship with alcohol. Life in the cities and the frontier was frequently hard, the availability of potable water for drinking was questionable, and taverns and saloons were one of the few social places for gathering.[34] Large numbers of often rootless men were prime customers for the many drinking establishments. In addition, married men would often over-indulge when they were paid, then come home to their wife and children empty-handed. Alcohol was ubiquitous and took many forms. Having discovered (or inherited from European immigrants) the formulas for making hard liquor, humble farmers found that they could distill their corn crop into whiskey and sell their surplus production to others. Fruit farmers found it easy to make brandy out of their crops and hops, and barley could be readily transformed into beer. Many people made their own alcoholic beverages, and for most households it became a normal inexpensive drink, notwithstanding the consequences of excessive indulgence. The United States did not have a wine culture, but it did have a drinking culture, especially when it came to distilled spirits.

Recognizing the above issues had been the siren call of the temperance movement. Such often poorly organized groups did not seek the abolition of alcohol, merely a moderation in consumption. However, some religious communities did cast it in a devilish light – the road to despair and depravity. Strict rules against alcohol existed among some Protestant denominations, and, given their normative control in certain areas, to stray from such edicts was tantamount to formal ostracism. But other religious groups simply urged moderation, not abolition. Not surprisingly, however, it was women who were at the forefront of developing movements to constrain alcohol because they were the principal victims of its problems. When a husband spent most of his earnings on liquor at the end of the day or week, there was no money for food, and spousal abuse not infrequently followed. Furthermore, the subordinate position of women made it hard for them to escape the domestic violence that sometimes ensued. As the nineteenth century unfolded, a more formal temperance movement emerged, not just among religious groups but also those who saw the health problems associated with alcohol. Excessive alcohol consumption, it was argued, impaired worker productivity.

Wineries (as well as many wine drinkers) did not really think of themselves as part of the overall drinking problem. They argued that wine was a beverage of natural temperance, as opposed to intoxication, and therefore they shouldn't be included in the opprobrium directed at distillers. Apart from some pockets in southern California, there was little enthusiasm in the state for the prohibitionist sentiment, so many felt optimistic that this perilous-sounding movement would be thwarted. Not only would it have dire consequences for many businesses, it was also seen as an affront to an emerging lifestyle that invoked European sophistication that came with wine and food. Alas, their opposition was poorly organized as well as fraught with contradictions.[35] They failed to appreciate the rancor of the abolitionists, who were uncompromising in their views. And their options for fighting the growing movement were difficult. Even though wineries did not identify with the distilling industry, if they opposed abolition they would be associated with and deemed supportive of it. This could ultimately hurt their position by painting them with the same brush as the demon-drink groups. However, distancing themselves from distillers (and their nefarious ways) would make them sound like prohibitionists, which would in turn create its own set of problems. They were selling alcohol after all, even if they believed it was far less harmful socially than hard liquor.

In a somewhat concerted attempt to reconcile these contradictory approaches, winegrowers in California did form the California Grape Protective Association in 1908 to differentiate themselves from distillers and brewers, and to promote wine as a more moderate beverage. But their efforts did not bear fruit and opposition, particularly among female voters, remained trenchant. Their attempt at exemption thus failed, and the state's (and country's) development as a wine-producing and drinking place soon came to an abrupt halt.

Localized legislation had been introduced to end public drinking in the mid-century onward, but this was fragmentary and not always enforced. But when Congress allowed states to pass and enforce laws banning the sale of liquor in 1913 it gave local authorities de jure control over interstate commerce provisions, essentially undermining the commerce clauses of the constitution. From there to the proposition to ratify the Eighteenth Amendment that would prohibit the sale of liquor in 1919 was an easy path, eventually formalized in the Volstead Act of that same year. To all extents and purposes, alcohol consumption was now illegal.

Aside from the obvious restrictions that the legislation imposed on the wine industry (and wine drinkers), there was a sliver of hope for wineries. The

Volstead Act permitted household heads to make "fruit juices" in their own homes, since it was assumed the volumes produced would be sufficiently minimal to obviate any drunken excesses. The result was a home winemaking frenzy that Napa winegrowers were happy to supply with grapes and concentrate for such purpose. Even though alcohol levels in such homemade wines were meant to be low, enforcement was minimal at best. In addition, in a curious nod to the old missions, a certain amount of wine could be made by wineries for sacramental use. The aforementioned Eshcol facility was one that managed to survive through Prohibition by making such wine. Both of these clauses kept many Napa wineries in business, and, in fact, plantings increased during the next few years.

Since many producers realized demand for wine would not quickly dissipate with Prohibition, they initially made concentrate from grape juice and shipped it to markets, often on the East Coast. However, it seemed that discerning customers wanted to see the actual grapes rather than concentrate, so wineries started shipping grapes. This was always fraught with problems as grapes do not travel well, especially thin-skinned ones from which quality wines can be made. Since consumers wanted to make fruity, dry red table wines, the solution was to ship thick-skinned varietals such as Alicante Bouschet, Zinfandel, Petite Sirah, and Carignan, as they traveled well and made a deep red wine. Yields on Alicante were the highest, which pleased growers, plus it was easy to grow. But it was never the predominant varietal in California (or Napa), so more of it had to be planted. The result was that better-quality grapes were often ripped up and vineyards replanted with higher-yield ones. By 1926, forty percent of plantings in Napa were devoted to Alicante Bouschet, thirty percent to Petit Sirah, sixteen percent to Zinfandel, and thirteen percent to Carignan.[36]

It is not surprising that vineyard replantings and even new plantings increased in the early 1920s to meet this new demand. In fact, California vineyard acreage doubled between 1919 and 1926, from three to six hundred thousand, and grape supply increased by 125 percent (reflecting a shift to higher-yield varietals). In the initial years, the price per ton of grapes went from ten to one hundred dollars.[37] However, the shift toward high-yield varietals ran counter to the growing pre-Prohibition sentiment that Napa's focus should be on quality.

Notwithstanding the CWA's market prowess, some individual winegrowers had earlier begun to capitalize on the value added that came with the valley's brand development, and embraced quality that allowed them to sell their wine at a price premium. That strategy now came to an abrupt halt because the market for quality grapes had largely collapsed. Almost overnight, the attention turned toward varietals that could sustain growers now that a different business model was effectively imposed. Many growers shifted their production accordingly to one that best met the new quantity/transportability metric. As Sullivan aptly notes, "One of the negative outcomes of Prohibition in California was the virtual disappearance of almost all first-rate wine grapes, either ripped up or grafted to shipper varieties. Such grafting was particularly common in Napa."[38] Several decades of struggle to normatively impose a commitment toward quality over quantity, and the corresponding enhancement of Napa's brand value, dissipated virtually overnight.

The rapid increase in grape production eventually resulted in an oversupply by the mid-1920s. Coupled with the Great Depression, grape demand and prices subsequently collapsed, leaving growers with a massive surplus. Faced with no outlet for their product, they ripped up vineyards, and there was virtually no replanting in Napa in 1927. In the years that followed, many Napa wineries closed, as did large industrial wine operations in the valley, including those owned by the CWA. Of the 140 wineries in Napa at the end of the nineteenth century, few remained open. Phylloxera and then Prohibition crippled the industry, and those farmers in the valley who had persisted with fruit-and-nut crops rather than switch to grapes felt vindicated, since their market continued to expand.

Conclusion

It has often been said that, when it is easy to grow grapes, typically less effort goes into making the wine. In some respects, this is the problem that beset California (and even Napa) as the nineteenth century unfolded. The intent of the early missionaries was to make wine for sacramental purposes and provide a civilizing influence for the indigenous population through a combination of sedentary agriculture and spiritual activities. The wine, however, lacked the gravitas of that produced by their Cistercian and Benedictine ancestors in medieval Burgundy. Perhaps the Franciscans lacked the skill, commitment, and perseverance of their earlier counterparts although they had far less time to devote to experimentation as their time was cut short through secularization. Conceivably, the grapes they used – Mission grapes – did not lend themselves to the pinnacle of gustatory excellence that Pinot Noir and Chardonnay afforded. When their activities were curtailed and their land expropriated, they did not bequeath a notable wine culture to the new owners, merely the ability to easily grow a varietal that appeared to meet relatively undifferentiated needs. Mission grapes certainly had "authenticity," but they lacked the "place-making identity" that wine anthropologist Marion Demossier argues is necessary for a region to acquire a semblance of a wine culture.[39]

When a market for wine had emerged by the middle of the nineteenth century, it was structured around the activities of powerful merchants. They viewed wine as a low-quality commodity that could be blended and sold in bulk to markets in the rest of the United States. Mission grapes continued to be popular among wine growers because they met the yield requirements, were easy to grow, and could be blended with other varietals that they experimented with. Following recommendations from the agricultural specialists at universities, new varietals were slowly introduced. However, they were more likely to be adopted by wealthier newcomers to the industry whose resources cushioned them somewhat from possible failures. In fact, it was when those same specialists invoked the mantra of quality over quantity that it became apparent that northern California regions were capable of making quality wine from certain varietals. Doing this consistently, however, meant forgoing the easy market of commodity grape production that had emerged with the growth of powerful wine merchants in the San Francisco area. The latter were able to shape and structure the market in ways that were beneficial to them and some large wine growers, thus providing little incentive for change. When harvests were good it satisfied other growers, but their position was always subordinate in the supply chain and contingent upon them pursuing a high-yield production policy – that is, until there was an oversupply, and the inevitable falling prices exposed their vulnerability.

While the predominant market was predicated upon satisfying production quantity, a growing sentiment among experts continued to point out the potential benefits of a quality emphasis. Climatic and soil conditions seemed to favor a high-quality/low-yield paradigm with the resulting wine potentially commanding a price premium at market. Moderate institutional support endorsed this approach, and technical knowledge and skillsets among growers were improving, albeit often in informal settings. But it was resource-rich individuals who took heed of these new ideas and were willing to experiment with new grape varietals. Less risk averse than smaller growers and better able to develop an identity for their product that was based upon a regional identity (Napa), they were the pioneers in creating an embryonic place identity for their wine. But market channels continued to reflect extant practices that best suited merchants, which enabled the latter to retain the upper hand. The struggle to realize the quality potential was thus all too often subordinated to a market that favored a mass-produced and homogenized wine.

When Napa wines were finally afforded the recognition of their quality at the beginning of the twentieth century, and several good harvests reinforced this emerging reputation, the collapse of the industry with Prohibition brought these accomplishments to a virtual halt. Wine would continue to be made during the "dry" years, but it came from grapes that quality producers (and experts) had assiduously been trying to relinquish. A partially formed wine culture in the United States had yet to be enthusiastic about California wine, and when quality was introduced into the debate, that same culture favored Old World producers with brand recognition. Napa's potential breakthrough into this rarified oenological atmosphere collapsed when it became apparent that surviving Prohibition required making a simple wine to meet basic pleasures. Others in California could do that just as well as Napa, and probably cheaper.

Notes

[1] Thomas Pinney, *A History of Wine in America; From the Beginnings to Prohibition* (Berkeley: University of California Press, 1989), 238.
[2] Charles Sullivan, *Napa Wine: A History* (San Francisco: The Wine Appreciation Guild, 2008), 9.
[3] Pinney, *A History of Wine in America*, 241.
[4] Sullivan, *Napa Wine*, 20.
[5] Ibid., 22.
[6] Pinney, *A History of Wine in America*, 241.
[7] Ibid., 256.
[8] These included Frontignan, Alexandria, Flame Tokay, and Catawba – the latter grown on the East Coast and in the Midwest.
[9] Pinney, *A History of Wine in America*, 262–3.
[10] "An Act to provide for the better encouragement of the culture of the Vine and the Olive," *Statutes of California*, tenth session, Sacramento, 1859, p. 210. See also Leon D. Adams, *The Wines of America* (Boston: Houghton Mifflin, 1973), 23.
[11] Vincent P. Carosso, *The California Wine Industry, 1830–1895* (Berkeley: University of California Press, 1951), 50.
[12] Pinney, *A History of Wine in America*, 273.
[13] Quoted in Sullivan, *Napa Wine*, 33.
[14] Ibid., 46.
[15] James Simpson, *Creating Wine: The Emergence of a World Industry, 1840–1914* (Princeton: Princeton University Press, 2011), 202.
[16] Pinney, *A History of Wine in America*, 273.
[17] *Daily Alta California*, March 15, 1880.
[18] Sullivan, *Napa Wine*, 72.

[19] If a particular vintage did not appear to meet desired alcohol levels, sugar might be added to the unfermented grape must. Such a process continues to provoke controversy, but at that time it was seen as introducing an unnatural aspect to the winemaking process, and could conceivably impair the final quality.
[20] Sullivan, *Napa Wine*, 73.
[21] Herbert B. Leggett, *The Early History of Wine Production in California* (San Francisco: Wine Institute, 1941), 112.
[22] University of California College of Agriculture, *Report of the Viticultural Work during the Seasons 1887–93* (Sacramento, 1896), 3.
[23] Sullivan, *Napa Wine*, 74.
[24] Sullivan, *Napa Wine*, 146.
[25] Pinney, *A History of Wine in America*, 350.
[26] Sullivan, *Napa Wine*, 133.
[27] Sullivan, *Napa Wine*, 80.
[28] See Earnest Peninou and Sidney Greenleaf, *Winemaking in California: III. The California Wine Association* (San Francisco: Porpoise Bookshop, 1954).
[29] Pinney, *A History of Wine in America*, 357.
[30] Although prunes continued to have a greater acreage than grapes at the beginning of Prohibition. See Sullivan, *Napa Wine*, 160.
[31] For a fuller discussion of this program see *USDA Bulletin 209* (August 6, 1915).
[32] Sullivan, *Napa Wine*, 166.
[33] Simpson, *Creating Wine*, 213.
[34] For a general discussion of alcohol and society in the nineteenth century see Ian M. Taplin, *The Modern American Wine Industry* (London: Pickering and Chatto, 2011).
[35] Pinney, *A History of Wine in America*, 435.
[36] Sullivan, *Napa Wine*, 193.
[37] Pinney, *A History of Wine in America*, 438.
[38] Sullivan, *Napa Wine*, 193.
[39] See Marion Demossier, *Burgundy: The Global Story of Terroir* (New York: Berghahn, 2018), 53. Demossier develops an analysis of Burgundy's historical trajectory in which place, culture, history, soil, climate, and social conditions combine to create a powerful narrative of evolving excellence of the region's wines.

CHAPTER THREE

NUNC EST BIBENDUM: THE 1930S–1960S

When Prohibition ended in 1933 the Napa wine industry bore little resemblance to its early twentieth-century guise. Only fifty-three wineries had survived and most of them were run down, undercapitalized, only capable of producing wines that were barely drinkable, and with staff who were not well trained and in limited numbers. What grapes that were planted were often the wrong varietals for making quality wine. In fact, most went towards making bulk wine that was sold either in barrels or to out-of-state bottlers. Furthermore, the market for quality wine that some wineries had demonstrated they were capable of making in the years prior to Prohibition had disappeared as consumers fell back on Old World producers, especially the French. And then there was the Great Depression, which stymied sales of many consumer products, a situation made worse in the wine industry with the oversupply of low-quality grapes. The Second World War further depressed sales and made access to materials for winery regeneration difficult. It was a dire situation, which James Lapsley aptly captures:

> In the first years following Repeal, California winemakers were untrained and unscientific, often not understanding the process of fermentation, and they lacked the technology to control it even if they did. As a whole the California industry was primitive and backward. Winemakers worked with antiquated equipment in undercapitalized enterprises at a time of cut-throat competition. It was not an environment conducive to the pursuit of quality.[1]

These are hardly encouraging words, and attempting to reinvigorate the industry would indeed prove a daunting task. Much that had been learned in prior decades had been lost, and a dearth of skilled winemakers meant the introduction of new knowledge was limited at a time when it was most needed.

Despite the overwhelming sense of despair, several trends nonetheless proved auspicious, and gradually the industry began to rediscover its earlier potential. As in the past, newcomers with either viticultural knowledge or financial resources injected an almost didactic determination to resuscitate the sector. Their formidable enthusiasm became a paean to the evolving reputation of Napa, most conspicuously through their efforts to resuscitate and organize activities in the wine market. In the beginning their numbers were small, but gradually increased. They were also instrumental in establishing advocacy organizations that could coordinate the marketing of premium wines to an emerging consumer market. The latter's growth was slow and needed persuading to switch their preference from fortified wines to table wines. But, gradually, they became receptive. Finally, local universities became more forthright in articulating the need for better techniques and pushed for the planting of premium varietals. As they unambiguously pointed out, to make a fine wine required quality grapes.

Of the newcomers with winemaking experience who joined the industry in the 1930s, the most notable was technically trained winemaker and scientist André Tchelistcheff, who was hired by Georges de Latour at Beaulieu in 1938. He was the driving force behind the growth of informal winemaker groups who shared valuable information across a wide range of viticulture and oenology. There were also a few new owners with extensive financial resources and the requisite capital to address many of the technical problems that beset wineries, and their presence proved salutary. Their extensive investments, unbridled enthusiasm and willingness to adopt a longer-term view of profitability potential allowed the sector to yet again focus on quality and the requisite grape varietals that were best suited to achieving that potential. Finally, more systematic viticulture and oenology research came out of the universities, and this proved indispensable for understanding basic processes that could guarantee the production of a consistent-quality wine. Much of this was spearheaded by professors Albert Winkler and Maynard Amerine, who established experimental sites at UC Davis and, with almost missionary zeal, advocated cleanliness and dramatic improvements in the fermentation process, plus the need to plant better grape varieties in the place best suited to them.

These evolving changes began the transformation of Napa. It went from a place beset with bulk-wine production, limited market demand for the few quality producers, and a concentration of poor-quality grape varietals in the 1930s to a region where a handful of larger producers were able to consistently make a wine and sell it under a distinctive Napa brand by the 1960s. Much of this was facilitated by two important processes that generated a more dynamic market. The first was improved access to and sharing of requisite technical knowledge among winemakers. Second, an initially informal then subsequently more systematic framework of cooperation among many of the new winery owners emerged. The fact that the valley was a small geographical space also helped, as proximity and density encouraged interactions. The technical knowledge to make quality wines became more transparent and accessible, while the new owners attempted to organize the market to ensure that the reputational benefits of quality wine production could be sustained. Both would be crucial in placing the industry on a firmer quality and financial footing by the 1960s.

The Post-Prohibition Decade

When Prohibition was finally repealed at the end of 1933, a visitor to Napa would have thought that it was primarily a fruit-and-nut growing area, followed by eggs and dairy. In fact, incomes from grapes accounted for less than thirty percent of farm revenue throughout the 1930s.[2] This was clearly not the wine country that many in the early decade of the century had envisaged and even begun to realize. In fact, it seemed that if you were interested in a commercially viable agricultural endeavor in the valley you became a prune farmer – the most valuable crop until the mid-1940s.

In the wineries that survived, some good-quality grapes existed, such as Zinfandel and Petite Sirah, but there was too much Alicante Bouschet (that had been planted during Prohibition for shipment to the East Coast) and very little Cabernet Sauvignon or quality white-wine grapes. Even more auspicious were consumer tastes. Interest in wine had not necessarily diminished during the "dry" years, but taste had shifted to sweet wines or coarse homemade ones. The former coincided with a growing availability of sweet non-alcoholic beverages such as soda pop; the latter because large quantities of poor-quality grapes continued to be shipped east through the 1930s. The few consumers with discerning tastes appeared to prefer French wines, but their numbers had not grown much over the past few decades. Thomas Pinney aptly sums up the general situation when he states, "the seemingly irresistible tendency towards the large scale, combined with the lack of a demanding critical market, made it almost inevitable that California winemaking in the first years after Repeal knew much about quantity and very little about quality."[3] Since most wine was blended and shipped in bulk, there was little incentive for wineries to spend much time on varietal labeling.

Given the apparently limited demand for quality wine, it is not surprising that many growers continued to focus on the production of basic table wine, shipped in bulk to bottlers on the East Coast. But in doing this they faced growing competition from central-valley growers, whose yields were much higher. Several University of California studies during the 1930s showed that yields of less than five tons per acre would not be profitable for growers.[4] Dry farmed (unirrigated) Napa vineyards at that time were producing yields of two-and-a-half to three tons per acre – a clear indication of the financial difficulty growers in the valley faced if they tried to compete with bulk growers. Yet, the expense and uncertainty of ripping up vineyards to plant better-quality varietals during the height of the Great Depression were not for the faint of heart. And, absent the demand for a quality product, it is not surprising that many growers simply muddled through, making wine as cheaply as possible regardless of the quality. Not surprisingly, some wine was of inferior quality, and in certain instances technically not even wine at all.[5] Napa's early embryonic reputation was clearly sullied.

A few of the larger wineries that had managed to survive through Prohibition were, however, beginning to make a comeback, and gradually transformed from making bulk wines to better-quality table wine. Their owners remained wealthy enough to continue operations and make the necessary capital investments, or they had the fortitude and discipline to explore alternative production methods. Beaulieu, Inglenook, Beringer, and Larkmead were the four leaders at this time, soon to be joined by the Christian Brothers (at their Mont La Salle winery). Despite low prices and weak demand, they planted more Cabernet Sauvignon vines, anticipating that this would be a premium varietal, production of which could be initially subsidized by their bulk-wine production. The owners of these wineries would meet informally to discuss ways of augmenting the reputation of their premium wines, and agreed that varietal labeling and clearly designating the Napa origin would be beneficial. If they produced quality wines they needed to sell them at a premium price, and consumers needed to be unambiguously clear about the quality of the product if they were to pay that price. Promoting Napa as a place from which quality wines came would not be easy, given the tendency for the shipment of marginal wines, so they needed to address the perennial problem of reconciling their quality production capabilities with public perceptions that stifled market demand.

A meeting of almost fifty winery representatives was held in San Francisco in 1934 to discuss a wide range of issues, such as distribution, simplification of regulations, and how to promote wine consumption among Americans. It morphed into a trade organization and became the Wine Institute (WI) when thirty-two wineries signed on. Its membership gradually expanded, but its focus was more industrywide, although it did address definitions of quality and identity for California wine. One of its successes was in shaping the Federal Alcohol Administration's 1935 standards report.[6] The WI operating mandate was broad, but its focus upon improving quality was beneficial to Napa producers. It would sponsor tastings, advocate for clear labeling of California wines, and assiduously promote wine consumption. A more focused group of winery owners in Napa would eventually seek to further differentiate premium wines from others produced in the state, and attempted to move beyond what the WI was doing. Notwithstanding such individual efforts, the WI did much to establish the credibility of California wine, and in doing so gradually changed consumer attitudes and prejudices toward it.

The nascent brand awareness that emerged from the institute's activities was further aided by a number of good harvests in the late 1930s. Additionally, some wineries were benefiting from their adoption of better viticulture and oenology techniques that the university was disseminating. That, and the arrival of winemakers whose chemistry background injected a much-needed shot of technical rigor into what was often still a haphazard and unsystematic method of winemaking, brought even more professionalism to the industry.

Fig. 4.1. Winery ca. 1930s

Source: author's sketch

During the 1930s, professors at UC Berkeley and the growing UC Davis campus extolled the need for an overhaul in winemaking practices. They also argued that varietal planting should be made in more appropriate places, i.e. growing cooler-climate grapes in the southern part of the valley around Carneros, and warmer ones in the mid and northern parts. Institutional support had come from universities in the late nineteenth century, but this was of a different magnitude. Often through the force of personality alone from key individuals, the post-Prohibition pattern involved more systematic extension activities. For example, at the University of California, W. V. Cruess was forthright in arguing that even the best current winemaking methods could produce wines of consistent quality, but were typically lacking in distinction. Others produced what he claimed were deplorable wines.[7] He went on to say that if better wines were to be made in California, better grape varietals were needed, practices of picking overripe grapes should cease, overall cleanliness in the winery and in transportation needed massive improvement, and the practice of trying to market wines that were spoiled should stop.[8] It was clear that much work needed to be done.

Two other professors, Amerine and Winkler, traveled throughout Napa highlighting the often-primitive techniques and unsanitary conditions that they found there. They visited many growers, extolling the virtues of new techniques that would, they claimed, maximize the potential of their vineyards. Premium-quality wine could be produced, they forcefully argued, but not unless a significant overhaul was implemented, even at stellar sites such as Beaulieu and Inglenook. Fermentation techniques needed changing, especially the control of secondary fermentations, pure-yeast strains introduced, and the elimination of the use of overripe grapes. These were just some of the modifications they advocated, yet almost all necessitated a rethinking of what grapes should be grown and how wine should be made. Along with other university scientists, they argued that improving the overall technical competence of winemakers could result in the production of good-quality wines on a consistent basis.[9] To not do so would be egregious, significantly impairing Napa's ability to transcend mediocrity and the cult of bulk-wine production.

Their optimism was laudable, but inevitably many growers were less than enthusiastic when hearing such complaints and opinions. The professors' ideas did find some traction (and enhanced credibility) when written up by wine-trade journals in articles that proffered extensive practical advice on how to improve quality.[10] In fact, the broad dissemination of such advice proved crucial in lending legitimacy to ideas put forward by academics. But unfortunately, many growers lacked sufficient capital to implement these changes, while others were deficient in the requisite skillsets to make them work. Furthermore, absent any significant benchmarking, it was easy for wineries to ignore the university specialists and continue to believe in the supposed quality of their own wines, even though objective evaluations might have proved otherwise. Either that or their resource constraints prevented them from contemplating the stark reality of what their product was like. As long as they were able to sell their wine they were content to let the market be the ultimate arbiter and ignore the noisy chatter from academic scientists. Not surprisingly, some left the industry, unable to either revive old properties or successfully start with the varietals that the specialists were recommending. It wasn't a case of obduracy, merely a lack of resources. On top of all this there remained uncertainty as to whether there was an actual sustainable market for premium grapes, meaning that the risk for under-resourced growers was often too great. Such recalcitrance was not necessarily the norm, but it was sufficiently widespread to impair the overall credibility of Napa's slow embrace of quality.

Embracing Change – Again

Of the 811 growers in the valley by the late 1930s, most were small operators, often unwilling to take a chance with planting higher-quality varietals. However, there were a few better-resourced wineries that recognized the potential for premium varietals, if nothing else as a way to avoid the destructive competition from bulk-wine producers. One such owner was George de Latour at Beaulieu. He had been around long enough to recognize that the future lay in premium varietals and not in bulk wines that would never be competitive with the growing production from the central valley. He hired a wine chemist, Leon Bonnet, from UC Berkeley's Agricultural Department in 1936, and together they set about crafting a Cabernet Sauvignon from his best grapes. In 1938 he then hired another first-rate chemist with a wine-industry background, André Tchelistcheff. This proved to be a seminal appointment.

Tchelistcheff was a thirty-six-year-old Russian who had acquired winemaking experience working in France and elsewhere in Europe. He was a researcher but also had an extensive practical background in both the vineyard and the cellar. He proved to be exactly the right person to apply most of the ideas and principles that academics at UC Davis were advocating since he had the practical experience and legitimacy that some of the academics lacked. He went about systematically introducing innovative techniques in the winery, from the introduction of cooler fermentation to a more controlled use of sulfur dioxide. In many respects he was uncompromising and a perfectionist in his routines, but the finished product proved to be some of the best wine

Napa had ever produced. When de Latour released BV's first private Reserve Cabernet Sauvignon in 1940, Tchelistcheff was the catalyst behind what would become recognized as a very good wine.

There were also others who were new owners who managed to prosper by embracing the growing body of knowledge that was being shared by the successful wineries. For example, Louis Martini was the first to introduce mechanical refrigeration at his St Helena winery, which he founded in 1933. This was an unambiguous attempt to heed the emerging advice on the benefits of cooler fermentation techniques. Other individuals, some with significant capital resources, shared a determination upon entering the industry to make great wine, and were purposeful in their adoption of new techniques. These included Cesar Mondavi, who acquired the Sunny St Helena winery in 1934 and eventually, along with his sons Robert and Peter, the Charles Krug winery in 1943; retired Southern Californian businessman Albert Ahern, who purchased the Lombardi winery in 1940 and renamed it Freemark Abbey; and Louis Martini who in 1942 sold his mass-market winery/distillery operation in Kingsburg and bought two hundred acres in Carneros to focus on the production of premium wines.

Another notable pioneer was Gustave Niebaum's great nephew John Daniel Jr at Inglenook who, together with winemaker George Deuer, focused on the production of estate-bottled wine that met rigid quality standards. As *Wine Spectator*'s James Laube commented, "for that amazing 31 year stretch – 1933 to 1964 – Inglenook compiled a collection of Cabernets that stand up favorably to the best red wines on earth; nearly all of these Inglenook wines were made under Daniel's inspired leadership."[11] He refused to sell wines under the Inglenook label that he felt were not of an appropriate quality because it would damage his brand. Such wines were "declassified" and consigned to bulk sales, with the result that in certain years (for example 1945 and 1947) there would be no Inglenook Cabernet Sauvignon. Such decisions were costly and unprofitable, and did eventually force Daniel to sell the winery in 1964 when he faced severe financial hard times, but he was renowned for his unwavering commitment to quality. As Charles Sullivan notes, "No one did more than John Daniel to achieve and maintain the high level of quality and the attendant prestige that Napa Valley premium wine received in the fifties."[12] Such praise was echoed by Tchelistcheff, who said Daniel's commitment to quality standards was not just remarkable, but "unbelievable."[13]

Each of the above charted growth through a commitment to quality, gradually eschewing the bulk-wine paradigm to which others continued to adhere. Also of note was the 1943 purchase of the 527-acre To Kalon vineyard near Oakville by San Francisco real-estate magnate Martin Stelling Jr. He immediately sold ninety acres to Beaulieu and then acquired an adjacent ranch and planted vines, but made no wine prior to his untimely death in 1950. This property eventually proved to be one of the iconic vineyard sites in the valley, producing some of Napa's most notable wines.

Wine consumption, both in California and nationally, was gradually increasing by the late 1930s, and prices were steadily rising. But there remained considerable price volatility, and the bulk of wine was of inferior quality and often fortified. Much of this wine came from large growers, often in the central valley were high yields and low production costs gave that group a distinct competitive advantage. Even as late as 1967, fortified wines outsold table wines, suggesting the continued preference of the American consumer for something sweet. This did not bode well for advocates of premium wine. Yet, at the 1940 State Fair in California, a more rigorous blind tasting awarded half the silver and gold medals for dry table wines to six Napa wineries. Clearly it was possible to make quality wines, even if the American wine-drinking public failed to appreciate them. If such attitudes were to change, however, wineries needed to find ways of communicating their capabilities to skeptical consumers, and not just individually, but with a collective voice.

Nascent Networks

Napa was small enough to facilitate regular social interaction among owners, and this proved auspicious for the creation of informal gatherings along the lines of what Tchelistcheff had been encouraging. In 1943, Martini was the force behind an informal group of Napa vintners that met occasionally over lunch at St Helena restaurants to discuss issues such as government price controls on wine, labor shortages, and the limited availability of railcars for shipping to the East Coast. Since these were pressing concerns for each of the members, competition was temporarily put aside.

In many respects, the group was similar to the nineteenth-century St. Helena Vinicultural club in that it provided an opportunity for owners to meet and share information, exchange ideas, and collectively promote the status of Napa wines. But in this iteration it became more formalized when, in 1944, seven of the most prominent vintners drafted an agreement of association and the Napa Valley Vintners Association (NVVA) was created as a formal trade organization.[14] This group proved an important advocacy vehicle and collective voice for winegrowers in the valley as well as a forum in which new ideas could be discussed. It is precisely this sort of collective organizational learning that would prove important for the adoption of new techniques as well as deal with production problems and reinforce a commitment to quality wine production. Whether or not the members were good friends, as some in the press had claimed,[15] they were nonetheless competitors who realized that collectively they could be more efficient at achieving their goals than acting as individuals.

The Second World War was a mixed blessing for Napa winegrowers. On the one hand they faced shortages of materials much needed in the winery as well as access to railroad cars to ship wine eastward. Ironically, because it became difficult for them to ship their wine in railcars, they resorted to bottling at the winery – something that had rarely been done in the past. Somewhat fortuitously, this gave them an opportunity to better brand their product and also reduce the likelihood of spoilage that often occurred with bulk shipments. It also gave them a way around price controls that had been introduced as part of wartime economic measures. High-quality bottled wines could now be more easily differentiated from bulk wines, and a price premium charged for the product. The quality versus quantity debate was not resolved, but this was exactly the sort of stimulus that foreclosed any dissimulation over the efficacy of premium wines. An emerging bifurcated market would grow more distinct in subsequent years, even encouraging some California bulk producers to become bottlers and build brands.[16]

During the war, the government required that many grapes be dehydrated and made into raisins for troop rations. This typically applied to the low-quality bulk-wine grapes, dramatically reducing the supply of grapes overall and a subsequent rise in market prices that benefitted the quality producers. For example, the price of grapes went from fifteen dollars a ton to fifty dollars during the war, significantly enhancing the profitability of growers. This proved beneficial to Napa growers, who had often struggled to compete with lower-cost bulk producers.

Wine consumption increased during the war years, and rising grape prices generated profits for wineries that many then invested in better equipment and procedures that further improved the quality of wine. Although the market overcorrected in 1947 when government intervention was replaced by normal market forces and prices fell, Napa was emerging as place where quality wine was more consistently produced, and increasingly recognized as such. The change was notable. As Lapsley comments, "the boom from 1943–1946 had proved short lived, but it was nonetheless pivotal to the future of California wine. In four short years, the California wine industry had changed from production of a bulk commodity to a pre-dominantly brand oriented business."[17] Those prewar years where quality producers struggled to compete with low-cost bulk producers on the basis of price had been replaced by a market that was developing an appreciation for quality wines, and distribution channels changed to ease bottled shipments. Restaurants, wine merchants, and hotels were selling more wines by their varietal label.[18] It might have been a temporary phase of post-war euphoria, but it signaled the benefits of premium versus bulk wines and was increasingly advantageous to Napa.

Fig. 4.2. Winery ca. 1940s

Source: author's sketch

Embracing Quality

Having finally recognized the profitability potential of premium wines and understood what needed to be consistently done in the winery to make this possible, winegrowers still deliberated on what varietals to plant. The old standby grapes such as Alicante Bouschet, Palomino, and even Mission had gradually been replaced with Zinfandel, Petite Sirah, Carignane, and Gamay – standard varietals that produced solid if not always exceptional wines. But there was minimal planting of Cabernet Sauvignon, Pinot Noir, and Chardonnay – varietals that were most likely to demonstrate Napa's full potential as a premium wine-producing area. But at that time there was still uncertainty among growers about varietals, despite continued exhortations from some winemakers as well as university specialists.

At UC Davis, professors Amerine and Winkler produced a comprehensive report in 1944 that examined quality issues, susceptibility to disease, the inherent vigor of vines, and resistance to frost, to name some of the most prominent.[19] This report contained systematic recommendations which were taken up by some of the prominent winegrowers. Meanwhile, informal technical groups among winemakers were meeting to discuss basic operational issues. Inspired by Tchelistcheff, who had created his Napa Valley Enological Research Laboratory in 1945, the Napa Valley Wine Technical Group (NVWTG) was formally announced in 1947.[20] Its members were winemakers, many with extensive industry experience or university training, and the aim was to provide a forum for the discussion of quality improvement and basic winemaking techniques.

The exchange of information derived from scientific studies was a crucial component of the group's meetings and included a range of issues such as the ongoing concerns over fermentation practices and yeast preferences. Most importantly, this group shared tacit as well as innovative technical knowledge among individuals with winemaking credentials – precisely the sort of collective operational knowledge transfer that is crucial for an industry to overcome the liability of newness (in this case, the absence of widely acknowledged and accepted techniques). Such discussions took a systematic, empirical approach to problem solving in order to discern best practices. Meetings were regular, albeit often in informal settings, and wines were tasted and critical opinions proffered. An informal benchmarking ensued that would be critical to the evolving understanding of which wines the region could most successfully produce. This growing professionalism among winemakers would eventually lead to the foundation of the American Society of Enologists in 1950, which firmly placed the peer-review process as the centerpiece of empirical science. As James Lapsley states, "the Napa Valley Wine Technical Group helped translate that science into practice and was a major force in improving Napa wines following the war."[21]

Consumption and Production Trends

Wine consumption continued to increase gradually during the 1950s and Americans were shifting from sweet to dry table wines. Consumption of quality wines was also growing, and the general post-war prosperity created an incipient wine culture in which wine was seen as an accompaniment to food rather than a desert type or standalone beverage. This was particularly evident in premium-wine consumption at restaurants, especially in San Francisco.[22] People were traveling and developing more sophisticated lifestyles, and via the media were exposed to a wider range of leisure activities. However, only about a third of Americans drank wine, and for many of them it was not more than once a week. The lack of a broad market for wine merely exacerbated other persistent problems such as periodic overproduction followed by market-price collapse.[23] It's perhaps not surprising then that there were still only thirty-five wineries in Napa in 1950. Many of the smaller operations had closed, as had the two big cooperatives (Napa Valley Cooperative Winery and the St Helena Cooperative Winery) that for years had been responsible for processing most of the valley's production.

In Napa, new plantings were nonetheless increasing and perhaps one significant trend was the gradual replacement of hitherto valuable prunes with grapes. Prune acreage fell from 9,000 to 7,651 acres in five years (1956–61), and declined rapidly thereafter to less than two thousand acres by 1971.[24] The agricultural focus of the valley during the 1950s was finally shifting to grapes, and some of the new plantings were of the better-quality varietals. Even though grape prices remained volatile, this nonetheless suggests that more people were recognizing the economic potential of making wine as opposed to basic fruit-and-nut farming. A good example of this, and of the growing involvement of UC Davis academics, was the new planting of Cabernet Sauvignon grapes on what had hitherto been a prune farm near Oakville. In 1959 the Rhodes family, following the advice of Professor Winkler, planted on a forty-two-acre lot they owned.[25] They subsequently sold their ranch to Tom May, who added more Cabernet plantings on an adjacent property and named it Martha's Vineyard after his wife. This would emerge in subsequent decades as one of the iconic Cabernet vineyard sites in the valley.

Yet, despite these changes, there were few consistent-quality producers in California, with approximately ten percent of the three hundred bonded wineries in the state producing what could be considered premium wine. While most of these were in Napa, even that area remained partially mired in bulk production. The best-known wineries and those that were increasingly recognized by the press as producing premium wines often relied on two production lines – one for their premium or estate wines, and the other for volume production that came from grapes purchased elsewhere in the state. Although some attempted to produce only estate wines for which they could charge a higher price, most relied on the volume production of more generic wine to cover their costs. The quality of the latter was not necessarily inferior, it merely enabled them to blend and offer a popular-priced wine to a larger market. But as long as this operational mandate persisted, it would continue to dilute their brand reputation.

Those that did designate their wines as estate bottled were capitalizing upon the public appreciation of premium wines (albeit with quality often vaguely defined) that developed during the 1950s with new wine writers and books on wine. Wine was now seen less as the preserve of sophisticated elites or itinerant alcoholics and more as a routine beverage appreciated for its contribution to a balanced lifestyle. It was also losing its stigma as mainly an immigrant drink, an image that consigned it to the periphery of respectable society. Wine and food were written and talked about in more common parlance, and this captivated many who felt their lack of knowledge had been an impediment to their appreciation of the beverage.

Estate bottled then, as now, conferred higher status and a price premium. Generally, this came from the same group of wineries that were mentioned in discussions of California premium wine, and Napa continued to be the pre-eminent region.[26] Exactly how much premium wine was made is difficult to discern. However, Lapsley refers to how, starting in the 1940s, the Gomberg Report's use of tax reports to differentiate different categories of wine shipments was a fairly ingenious way of tracking sales when other reliable data was lacking.[27] They indicate that growth rates varied during the 1950s for the top producers, but took off for most wineries in the 1960s.

Efforts continued throughout the 1950s and into the 1960s to convince the American public to drink premium wines. But to do that, many of that same public had to be informed as to what exactly quality was. How did one define it in terms that could be understood by consumers who were often new to the product? Did a high price mean quality? Was an estate wine necessarily better than a generic one? Individual brands could continue to argue that their wine was premium, but it needed a more coherent collective marketing effort to make a broader convincing case. Some wine lovers still associated quality and premium with Old World, particularly French wine. Now that post-war production and shipments were resuming from Europe, the challenge for Napa producers to overcome this cultural bias was even more difficult.

Promotional Efforts

The NVVA increased their advocacy work during the 1950s and frequently invited influential press members and other notable figures to visit Napa and taste the wine. However, they realized that a more prominent national effort was needed, and in 1955 makers of premium wine in California formed the Premium Wine Producers in California Association (PWPCA). This was nominally a trade organization, but it also sought clarity on the definition of a premium wine, positioned California premium wines as an alternative to imported fine wine, and embarked on a publicity campaign to convince American consumers that wine was part of the "good life."[28] It also unambiguously emphasized the growing importance of the premium wine segment in California. Wine consultant Louis Gomberg[29] was hired to coordinate the efforts, and he immediately set about establishing tastings in major US cities. These gave the broader but also more affluent section of the public an opportunity to try a variety of California wines and compare them with premium wines from Europe. In most of these tastings, as well as at other evaluations, Napa wines garnered just over fifty percent of the top awards.[30] While ostensibly a California-wide initiative, in this quasi-objective setting consumers could appreciate exactly what the best Napa wines were like, and hopefully be made less cautious about buying them.

Given the growing consumption of such wines by the early 1960s, it is reasonable to infer that the campaign was successful. However, it did provoke the ire of non-premium producers such as Gallo who felt that their exclusion tarnished their reputation, even though they were unambiguously bulk producers. They and others claimed that the industry in general was subsidizing the premium segment. The PWPCA response was to argue that the campaign benefitted everyone as it more firmly placed wine consumption as part of everyday American life. Such a claim was not entirely disingenuous.

Gallo had become a force in the industry since they built their winery in Modesto in 1933. By virtue of their volume and the high-yield production capability of central-valley grapes they became one of the leaders in the California wine trade by the end of the 1930s. Their reputation rested on their ability to supply large amounts of fortified wine to East Coast markets. Like many of the other big wineries, their focus remained on bulk wines, and their operational scale gave them a significant competitive advantage in this market. In 1938 they began bottling and labeling wine under their own name, and were extremely successful in building a brand for this wine.[31] In many respects, theirs was the easy-drinking wine that many Americans were familiar with – it did not require a sophisticated palate or a middle-class income to afford bottles. Furthermore, they were consummate marketeers and salesmen.

It is possible that by introducing consumers to their wine they might have encouraged some to trade up to premium wines – precisely the sort of activity that we now know happens. An interesting speculation at least! Their continued market power nonetheless meant that the quantity/quality debate would not be evanescent anytime soon. If Napa was to break free from the bulk-wine stigma it needed to more forcefully differentiate itself from others whose commitment to premium wine was more tenuous.

The 1960s Transformations

As with other periods in Napa's history, certain individuals played a key role in the 1960s as the growth of premium wine continued. Of particular importance were the Mondavi family who had built up operations at the Sunny St Helena winery, and then in 1943 purchased the Charles Krug winery. Cesar Mondavi's sons Robert and Peter assumed more active roles as production increased at the Charles Krug winery, where the focus was on better-quality wine (the bulk-wine production continued at the Sunny St Helena facility). Unfortunately, the two brothers did not share the same operational vision for the winery, with Robert more forcefully arguing that standards could be improved to make even better-quality wine. The resulting acrimony led to Robert's dismissal, whereupon he borrowed money to start his own winery. His hacienda-ranch style and now-iconic Mondavi winery near Rutherford crushed its first grapes in 1966. It was the first big new winery built since Martini's in 1934.

Robert's aim was to combine the best traditional practices with modern technology and use quality grapes to make a premium wine that would compete with the best from Europe.[32] To accomplish this he hired a new winemaker, Warren Winiarski, who was making a reputation for his technical skill at Souverain Cellars working under Lee Stewart. Winiarski's scrupulous attention to detail and his practical skills attracted him to Mondavi, and Mondavi's willingness to embrace the best innovative technology available ensured Winiarski's enthusiasm to work with him. When Winiarski left to start his own winery several years later, Mondavi hired another up-and-coming winemaking star, Mike Grgich, who had worked under Tchelistcheff at Beaulieu and gave him free rein to implement more technological innovations. He was responsible for further improving the quality of wine at Mondavi, and affirmation of this came in a 1969 blind tasting of the best California Cabernets when Mondavi's wine came top.

When Grgich left in 1972 to work for Chateau Montelena – where he was to further embellish his reputation and that of another emerging winery – it became apparent that Mondavi's commitment to excellence was unwavering since he had nurtured two of the best new winemakers in California, replete as they were with the newest technical skills and a passion for making fine wine. It also helped that others in the valley were embracing many of the new techniques that the university continued to disseminate. In addition to the formal associational groups that had been established, close geographical proximity continued to encourage tacit information sharing among owner and winemaker groups. This incipient clustering and cooperation helped consolidate best practices and formalized a culture of innovation.

It was during the 1960s that the gradual shift to better-quality varietal grapes began, in part a response to the encouragement of university specialists, but also necessary because old vines needed replacing. Since many of the vines had been planted forty to fifty years earlier, they were bearing less fruit. With a gradual rise in grape prices, especially for quality fruit, wine growers did what farmers everywhere do, and with some of their profits increased production by planting more quality varietal vines. Providing there was a continued growth for premium wines, this strategy proved efficacious. In fact, as demand grew and prices increased, it also encouraged wineries to purchase independent vineyards from which they had hitherto bought grapes. Giving them greater control over plantings and garnering a regular supply of precisely the grapes that they wanted without incurring price variations, wineries could eliminate some of the market uncertainty. It also enabled them to introduce requisite quality controls that were not possible when they were buying grapes. Examples of such a trend were the Christian Brothers' purchase of the 520-acre Forni Vineyard in 1956 and Mondavi's acquisition of the 500-acre To Kalon vineyard in 1962. Beaulieu and Martini also purchased properties in Carneros, and Beringer expanded into Knights Valley.[33] The result was a gradual takeover of many smaller properties by the end of the 1960s, signifying a structural change in ownership patterns. Consequently, by 1970 there were more "estate" wineries and fewer growers without a winery in Napa.

It was also becoming increasingly evident that passion and skill alone were insufficient for commercially success. Wineries needed significant amounts of capital as well as the administrative skills to market and organize distribution. It was one thing to make a premium wine, but quite another to consistently sell it. Undercapitalized wineries struggled, whereas newcomers with extensive financial resources, including corporate ventures, were increasingly attracted to the valley. For some, it was a lifestyle choice. Money that had been made elsewhere enabled them to "buy into" the wine industry and become immersed in the presumed Arcadian pleasures of agricultural life. This was not a new trend, but it did gain traction during the 1960s when Napa's image was on the ascendency. Examples of this are Donn Chappellet's 110-acre vineyard started in 1969 with money he made as owner of a successful food-service company, and southern California businessman Michael Robbins's purchase of a property near St Helena in 1962 where he planted a vineyard he named Spring Mountain (bonded in 1968).

The attraction of owning a winery in the up-and-coming Napa Valley extended to corporations. Corporate ventures – such as the United Vintners purchase of Inglenook in 1964, Heublein's acquisition of Beaulieu in 1969, and Pillsbury's purchase of Souverain (Lee Stewart's old winery on Howell Mountain) in 1972 – brought significant capital to wineries, but not always the long-term patience needed in the wine industry. Under corporate ownership, shareholder pressure for short-term gains sometimes proved incompatible with the basic economics of wine. The result was that some corporations lost interest and sold off their investments. For example, after three years, Pillsbury decided to offload their winery since they were losing six million dollars a year – clearly, wine proved far less profitable than their Burger King chain.[34]

One of the most notable events during this decade was the increase in premium varietal plantings that are now most associated with Napa. In 1961, more than fifty percent of acreage was devoted to Petite Sirah (1,748 acres), Zinfandel (949), and Carignane (819). Cabernet Sauvignon had only 387 acres planted (approximately six percent of the valley's total acreage), Pinot Noir 166, and Chardonnay 60. There was no comparable dominance of white grapes in the more than fifteen varietals planted. However, the ensuing decade saw a dramatic increase in plantings of premium varietals. By 1973, Cabernet Sauvignon acreage was 2,432, Pinot Noir 1,013 and Chardonnay 785 – compare this to Zinfandel (747), Petite Sirah (1153), and Carignane (534). The latter varietals had not disappeared, but since the total bearing acreage for red varietals had increased by almost a third (from 6,310 to 8,190 acres), the increase appeared to have come from new plantings as well as replantings. As the price of premium varietals increased and the quality of wine from such varietals continued to improve following better and more professionalized winemaking, growers and wineries shifted their production accordingly. This marks the beginning of the valley's embrace of the three varietals that would subsequently seal its fame.

Many of these trends can be understood by examining market-price differentials between premium and bulk wines. Harvest prices for standard (bulk) wines in 1961 were 120 dollars a ton compared with two to four hundred for varietals. By 1967 these figures were respectively 140 versus 275 dollars a ton. In 1960, *Wines and Vines* reported more and more interest among Napa growers in replanting with fine-wine varietals that included Cabernet Sauvignon and Pinot Noir as well as Alsatian-style whites (Riesling and Gewürztraminer).[35] Four years later, the principal interest was in Cabernet Sauvignon and Chardonnay.[36] Again, university specialists, especially those at UC Davis, were instrumental in providing regular information that would enhance grape growing and winemaking with these varietals. They summarized the scientific and technical reports that were increasingly published in academic journals such as the *American Journal of Enology*, and *Wines and Vines* was proving to be a popular source for the implementation of simple practical techniques. Unlike in the past where such innovation had been resisted or unaffordable, most wine growers and wineries were now willing (and mainly able) to experiment if it meant a better-quality wine. Finally, a market for their product was emerging that would make changing worthwhile.

Conclusion

The decade after Prohibition was a dormant one for the wine industry in general, and Napa in particular. The majority of wines made and sold were fortified and clearly very different to what aficionados deemed an acceptable beverage. It was not until the 1940s that some of the older wineries were successfully resurrected and began realizing their earlier potential as newcomers with the requisite financial resources took charge. During this time, growing institutional support, primarily from university academics, provided knowledge of the scientific and technical skills that could enhance grape growing and winemaking. The arrival of skilled winemakers from the late 1930s onwards gave a much-needed boost to the quality upgrading that was occurring at some properties, as well as being a stark reminder of how outdated and technologically backward many wineries still were. Napa was beginning to realize its potential to make premium wines, but to do so required capital investments and the professionalization of most practices. It also needed a viable market for its product.

By the 1950s, as more wineries were established and many independent growers bought out, a critical mass developed among owners that facilitated routine interactions. A local network structure had emerged that enabled the flow of information between key actors. Formal and informal groups of owners and winemakers met regularly (and independently) to share information, discuss advocacy strategies, and preserve the integrity of the growing Napa brand. This incipient clustering was crucial for knowledge dissemination, and in the case of owners provided an opportunity to order the market. It also engendered reputation building as individuals adhered to quality standards and reliability that were mutually enforced (what economist Diane Coyle refers to generically as "corporate goodwill").[37] Cooperation between businesses who are ostensibly competitors is a unique feature of an embryonic industry in a geographically specific region. It can be predicated on the mutual benefits derived from establishing accepted operating norms and the realization that, collectively, wineries could achieve a reputational status that would be more difficult for individuals acting alone. The growing legitimacy of the region out of its auspicious 1930s rebirth was testament to precisely this sort of collective organizational learning. Finally, these developments need to be seen in the context of growing institutional support for the industry, specifically the provision of public goods via university scientists whose endeavors promoted collective interests.

The creation of formal "trade" organizations to market quality wines and protect an incipient brand identity was also instrumental in shaping public opinion and nurturing consumer interest in wine. As more Americans became wine drinkers they learned to appreciate premium wines, aided in this evolving discernment by professional taste arbiters. The quality of wine was improving because better-quality grapes were used. And the technical competence of winemakers, together with transformations in the physical infrastructure of wineries, resulted in more consistent quality production. Despite these positive aspects, the industry was still a work in progress. Napa had emerged as the leader in quality production, but its reputation was far from established. It would take an acceleration of these trends and a very public international tasting in the 1970s to finally cement the early accomplishments.

Notes

[1] James T. Lapsley, *Bottled Poetry: Napa Winemaking from Prohibition to the Modern Era* (Berkeley: University of California Press, 1996), 40.
[2] Charles Sullivan, *Napa Wine: A History* (San Francisco: The Wine Appreciation Guild, 1994), 208.

[3] Thomas Pinney, *A History of Wine in America: From Prohibition to the Present* (Berkeley, University of California Press, 2005), 69.
[4] Lapsley, *Bottled Poetry*, 45.
[5] Sullivan, *Napa Wine*, 214.
[6] Pinney, *A History of Wine in America*, 99.
[7] Ibid., 86.
[8] William V. Cruess, "Suggestions for Increasing Consumer Interest," *Wines and Vines* (November 18, 1937), 6.
[9] Pinney, *A History of Wine in America*, 106.
[10] See, for example, *American Wine and Liquor Journal* (September 1941); also, the descriptions given in Albert J. Winkler, "Viticultural Research at the University of California, Davis, 1921–71," (1972) https://www.lib.berkeley.edu/libraries/bancroft-library.
[11] James Laube, "The Glory that was Inglenook," *Wine Spectator* (October 17, 2001).
[12] Sullivan, *Napa Wine*, 258.
[13] Tchelistscheff Oral History (1979), 112. Bancroft Library, Regional Oral History Office. https://www.lib.berkeley.edu/libraries/bancroft-library.
[14] The founding members were Fernande de Latour of Beaulieu Vineyard, Elmer Salmina of Larkmead, Charles Forni of the Napa Valley Co-op, Robert Mondavi of C. K. Mondavi and Sons, John Daniel Jr of Inglenook, Louis M. Martini, and Louis Stralla.
[15] See, for example, T. Eurch, "Lunch with the Brotherhood," *San Francisco Examiner* (October 13, 1974), magazine section.
[16] Lapsley, *Bottled Poetry*, 102.
[17] Ibid., 110.
[18] John Briscoe, *Crush: The Triumph of California Wine* (Reno: University of Nevada Press, 2018), 211.
[19] Maynard A. Amerine and Albert J. Winkler, "Composition and Quality of Musts and Wines of California Grapes," *Hilgardia* 15, no. 6 (1944).
[20] Lapsley, *Bottled Poetry*, 133–4.
[21] Ibid., 134.
[22] Briscoe, *Crush*, 215–6.
[23] Pinney, *A History of Wine in America*, 205.
[24] Sullivan, *Napa Wine*, 258.
[25] See P. E. Hiaring, "Martha's Vineyard," *Wines and Vines* (May 1979): 70–3.
[26] According to a 1948 book by R. L. Blazer titled *California's Best Wines,* thirteen wineries were listed, of which five were in Napa. The thirteen were: Almaden, Beaulieu, Concannon, Fountain Grove, Freemark Abbey, Inglenook, Korbel, Charles Krug, Los Amigos, Louis Martini, Novitiate of Los Gatos, Paul Masson, and Wente. See Lapsley, *Crush*, 141 for details.
[27] Lapsley, 142. Founded by Louis Gomberg in the late 1940s, Gomberg, Frederikson & Associates monitors shipments of California wine and imports and offers analysis of business trends in the wine industry.
[28] Lapsley, *Bottled Poetry*, 148–9.

[29] He used his columns in *Wine and Vines* to emphasize the growing popularity of premium wines.
[30] Sullivan, *Napa Wine*, 254.
[31] Pinney, *Bottled Poetry*, 193.
[32] Julia F. Siler, *The House of Mondavi: The Rise and Fall of an American Wine Dynasty* (New York: Gotham Books, 2007), 62–3.
[33] Lapsley, *Bottled Poetry*, 174.
[34] Sullivan, *Napa Wine*, 293.
[35] *Wines and Vines* (January 1960).
[36] J. Fiske, "Napa County," *Wines and Vines* (February 1965), 31.
[37] Diane Coyle, *Markets, State and People* (Princeton: Princeton University Press, 2020), 159.

CHAPTER FOUR

BEHOLD THE FUTURE, FORGET THE PAST

The 1970s were a pivotal time for Napa and its ascendant wine industry. More newcomers, often with plentiful resources and skills, were entering the valley, and alongside sustained corporate interest were continuing to revitalize winery operations. Quality was consistently improving as bulk-wine grapes were replaced with premium varietals such as Cabernet Sauvignon, Pinot Noir, and Chardonnay. Winemakers were better informed about technical procedures, and had the resources to experiment and a wider range of operational knowledge that systematized best practices. The universities, especially UC Davies, were continuing to extol the virtues of making premium wine, and were training new generations of viticulturalists and oenologists with a formidable range of scientific skills. The market for wine was also changing as demand for better quality accompanied an emerging American wine culture and increased wine consumption.

The 1960s had brought heightened economic prosperity, and, alongside larger numbers of women entering the labor force, saw a rise in median household income. A more affluent, consumerist lifestyle that accompanied middle-class prosperity was a departure from the previous decades of uncertainty that had contributed to price volatility and often depressed wine sales. People were frequently dining out and pursuing more ambitious leisure activities, and wine consumption was becoming integral to such pursuits. Supply-and-demand conditions were changing, and the premium domestic wine market developed a more ordered form and structure.

Entrepreneurs, Interlopers, and Corporations

By the end of the 1960s there were only eighteen wineries in operation in Napa, compared to thirty-five in 1950. In the decades after the Repeal of Prohibition, many smaller ones had sold out or closed, while some of the larger ones such as Inglenook faced financial difficulties and were purchased by corporations. It was increasingly clear that the switch to making premium wine was eminently plausible from an oenological point of view, but financing such an endeavor proved far more challenging. Furthermore, one had to sell the wine, which entailed marketing and distribution skills, adding to the requirements for a successful operation. Undercapitalization proved problematic, further compounded by high interest rates, often forcing owners to choose either compromising the ways they did things or selling their winery. Many were forced to do the latter, including the two big cooperatives, Napa Valley Cooperative Winery and the St Helena Cooperative Winery, both of which had processed much of the valley's grapes in previous decades.

Despite this bleak picture, owners such as Robert Mondavi succeeded in making a premium wine (although he did use some bulk sales through his Woodbridge brand to cross-subsidize such efforts) and continue to expand. He was the beneficiary of capital from the settlement with the family's Krug property, and this provided the cash he needed to hone his focus. He continued to extol the virtues of experimentation and information sharing, which he saw as indispensable for Napa to fully realize its potential as a premium wine-producing region. This almost missionary zeal was infectious, and others embraced the salience of developing collective organization and production knowledge.

Other newcomers succeeded because they too had the financial resources to make the requisite capital investments in the technology necessary to make a quality wine consistently. Good examples of this were Chappellet (1967), Sterling (1969), and Trefethen (1969). In some instances, old vineyards changed hands with the new owners revitalizing operations. Such was the case with the old Tubbs estate in Calistoga named Chateau Montelena. Founded in 1882 by Alfred Tubbs, whose fortune came from a Gold Rush-era rope business, by 1896 it was the seventh largest winery in Napa. After Prohibition it fell into disrepair and resembled a far-Eastern theme park rather than a winery. It was bought in 1968 when a group of investors spearheaded by Leland Pasich replanted the vineyards and focused production on premium varietals. They subsequently (in 1972) hired Mike Grgich, who had worked at Beaulieu and Mondavi, and gave him free rein to focus on quality.[1] The results were some impressive wines. Similarly, Brother Justin Meyer left Christian Brothers to become a secular winemaker when he partnered with Colorado oilman Raymond Duncan to form Silver Oaks Cellars in 1972, and focus on making a premium Cabernet Sauvignon. This pattern of an established Napa winemaker joining newcomers prepared to bring an infusion of capital occurred elsewhere during the 1970s.

By 1970 it was unambiguously clear that being a successful producer of premium wines requires considerable amounts of capital. The price of vineyard land was increasing in part because the valley came to be seen as a desirable place to live, and corporate interest in winery ownership became a way of diversifying portfolios. By the end of the 1970s it was not unusual to see Napa vineyard land sold at around twenty-five thousand dollars per acre. If one considers yields and grape prices, anything more than thirteen thousand per acre was not justifiable on economic terms. In other words, those buying such lands were not always the rational profit maximizers that economists like to reify but individuals whose interest in owning a winery was more of a lifestyle desire. Made possible by extensive financial resources from other businesses, this pattern would be repeated often in the decades to come. Ironically, however, their prospects of actually making money improved somewhat as wine consumption increased in the United States, and Napa wineries were able to capitalize on the trend by charging higher prices. What had started as an expensive hobby might eventually be more financially lucrative than their earlier intent.

The growth potential for wineries following increased consumer demand also piqued the interest of the financial sector. Increasing per-capita wine consumption might be a stimulus for vineyard acquisition among newcomers, and this was noted by leading banks such as Bank of America and Wells Fargo. Reports by the latter commented on the significant prospects of sustained growth in the California wine industry if this demand was to be met.[2] As a result of these optimistic projections, both banks were willing to loan money, and this fueled further investor interest in the industry.

Operating a vineyard was increasingly acknowledged to be expensive, but prevailing fiscal policies were something of a boon to those attracted by the lifestyle and potential financial yields. Tax law permitted write-offs for vineyard development which, together with other miscellaneous deductible expenses and the amortization of capital investments, made ownership less financially onerous.[3] Owner partnerships were particularly attracted to such terms as they simultaneously provided quasi-winery ownership status with positive returns and limited liability. This was the case for a group of investors headed by William Jaeger and Charles Carpy who bought Freemark Abbey in 1966. The winery had a storied history from its founding in the 1880s, but had been shuttered in 1959.[4] Upon their purchase, the group renovated the winery and tasting room, and focused on Cabernet Sauvignon and Chardonnay as premium varietals – a focus that they believed would deliver the best potential wines and the most reliable revenue stream.

Corporate interest in wineries also continued during this period. In the southern Carneros part of the valley, the French company Moët-Hennessy purchased a 550-acre ranch in 1972 to add to the two hundred they had acquired on Mt Veeder. Later that same year they purchased 350 acres of a dilapidated vineyard near Yountville. This was a significant foreign investment hitherto unseen in the valley. Their express aim was to make a sparkling wine in the same vein as Champagne, and they built a facility that could produce two hundred thousand cases a year (a figure they reached in 1983). Despite rising land prices in Napa, property was still cheap compared to the Champagne region of France, plus there was more of it. Relatively inexpensive land and its reasonably plentiful supply comprised the perfect cost proposition for their strategy of appealing to the growing wine-consumer market in the United States. Furthermore, the requisite grapes Chardonnay, Pinot Noir, and Pinot Blanc (eventually followed by Pinot Meunier) grew well in the area. Their aim was to create a sparkling wine that tasted like Champagne and could be sold at a price premium. Accordingly, they priced their wine above other California sparkling-wine producers such as Korbel, and the eventual name of Domaine Chandon was designed to confer a level of European sophistication that would justify the value-added proposition. The venture proved very successful, even though they were unable to label the wine as Champagne because it would contradict the exceptionalism that the French (and their parent company) had argued made their wines so unique. Nonetheless, Americans took to Chandon Brut and Chandon Blanc de Noir with great enthusiasm, as they have continued to do in subsequent decades.

Aside from partnerships and corporate groups discussed above, there were individuals who sought an escape from corporate America and saw in owning a winery the perfect alternative career path. As with their 1960s counterparts, they brought enthusiasm, commitment, and passion, but also financial resources that sustained their endeavors. Lou Gomberg referred to them as a new breed whose determination to succeed was all the more remarkable since they came from a non-wine-industry background. Such owners would, he claimed, propel California wine to "future greatness."[5] Many of these individuals were responsible for the rejuvenation and dramatic growth of wineries in Napa during the 1970s, with eighty-eight new ones established between 1967 and 1980. Furthermore, most were unambiguous in their focus on making quality, premium wine in smaller batches. Often referred to as "boutique" wineries, they personified "the search for excellence" that authors have categorized as the organizational impulse behind the region's continued embrace of quality-wine production.[6]

With vineyard acquisition came planting and replanting. Zinfandel and Petite Sirah were increasingly shunned for their rusticity, with more and more wineries planting Cabernet Sauvignon, followed by Chardonnay And Pinot Noir. By 1973 these three varietals constituted thirty percent of Napa vineyards, and that would double in the next ten years. They were seen as the best value-added products, and one gains a sense of this shift by looking at tonnage production and value. In 1960, Napa produced 23,776 tons valued at $2,211,168; by 1970, production was 43,673 and the value $8,055,000; and by 1973, it was 55,659 and $33,916,000. Tonnage more than doubled during this period, but grape value increased significantly (allowing for inflation). The grower price per ton went from $93 in 1960 to $184 in 1969 and $609 in 1973.[7] In 1967, grapes accounted for approximately one quarter of agricultural income in Napa, but by 1980 that figure was seventy-five percent. Vineyard acreage doubled from eleven to twenty-two thousand bearing acres during that same period.[8] The days of Napa's agricultural foundation resting on nuts and prunes (plums) was clearly on a downward trajectory.

Not all of the newcomers to Napa were aspiring vineyard owners, however. As the region's reputation grew so did interest from other parts of the country as people sought to relocate and enjoy a tranquil and dignified lifestyle surrounded by vineyards. Whatever rusticity the area had possessed in earlier times was gradually being replaced with a veneer of cultured sophistication built on a beverage associated with a more refined demeanor. The result was an area that had many of the benefits and charm of rural life without the negative connotations of backwardness. This is what made it particularly attractive to wealthy newcomers who wanted to escape from urban life without forgoing the essential trappings of culture. A similar trend south of San Francisco was transforming the Santa Clara area into a suburban community. With state population projections indicating a sustained growth in future decades, it was feared that Napa could follow suit and become increasingly suburbanized. Simmering discontent among many valley residents became more organized as these concerns bubbled to the surface.

A Rural Idyll Transformed

In the 1960s, Napa was a fairly tranquil place with a small railway that transported goods and people as far as Calistoga. St Helena had all of the characteristics of a small farm town surrounded by fields and vineyards. The specter of all of this changing arose when in 1966 the state discussed plans to build a freeway north of Yountville. The prospect of more traffic and homebuilding galvanized opposition among environmentalists and some vineyard owners who wanted to protect the agricultural nature of the valley. They in turn were opposed by developers and other farmers who endorsed the proposal as a way to develop the valley in line with their own pro-growth commercial interests. When another proposal to rezone part of the To Kalon vineyard to build cluster homes emerged it finally ignited the flame of opposition to further development.

Groups opposed to these developments coalesced around what was essentially a pro-agriculture policy that would change minimum zoning in unincorporated areas from one to forty acres. This was a systematic attempt to rebuff developers as it would make it very difficult to subdivide properties, effectively limiting land acquisition to agricultural endeavors. County Planning Commission meetings where these pro and anti-growth proposals were discussed were quite animated, to say the least.[9] Much also depended on the composition of the board and whether members were pro-growth or opposed to changes that would fundamentally alter the ecology of the valley. After extensive deliberation, the board finally voted to implement the raised minimum-zoning changes. This went into effect on November 11, 1968, and twenty-four thousand acres of agricultural land north of Napa were rezoned, whereby parcels could not be cut to less than twenty acres. In 1979, the minimum was raised to the originally proposed forty acres. The pro-growth group interests had been defeated, but the issue would remain contentious for decades to come.

The creation of what would become the Agricultural Preserve is recognized as a singular event in which local groups were able to shape public policy in ways that would preserve vineyard land. It was seen as a victory for winery owners who wanted to maintain the valley's agricultural ethos, although some who saw the prospects of heightened land value were not so enthralled. This was one of the first salvos fired in an ongoing simmering war over the essential nature of the valley. Its success demonstrated the power of many vineyard owners to resist commercial developments that would probably result in a rise in land prices. But it also prefigured subsequent discussions on tourist development, land-use practices, and the evolving character of the valley as a destination. This debate is ongoing to this day.

A Wine Culture Emerges

As Napa producers continued their emphasis on making premium wines and adopting better production techniques, so the quality improved. The more pervasive use of stainless-steel tanks and cold fermentation, together with better-quality grape varietals and more rigorous vineyard management that focused on quality rather than quantity harvesting, resulted in significantly better wines.[10] This was good news for the new consumers who were discovering wine as part of their newly affluent lifestyles.

The increasing prosperity of the 1960s and 1970s was associated with broad economic growth trends, wage increases amongst the middle and upper middle class, and a significant increase in female labor-force participation. As more women entered the workforce, the domestic culture changed. Rising median family incomes allowed people to travel more, eat out often, and sample activities that had hitherto been more exclusive. These were significant cultural changes as people embraced an experiential lifestyle that included foreign food and wine consumption. Wine was seen as a pleasurable beverage, lacking the drunkenness connotations of hard liquor and certainly more sophisticated than beer. Furthermore, the young "baby-boomer" generation viewed it as a way of marking their difference from their parents, and much of the increased wine consumption during the 1970s came from this under-thirty group.

Many authors including wine expert James Lapsley saw the heightened cosmopolitanism of American society in the 1960s as a turning point for the domestic wine market.[11] Increased affluence was part of the massive cultural changes that were occurring. Consumers were better informed about goods, individualism became more paramount, and people felt confident participating in the unfolding future of rampant materialism. They sought diversity and variety in what they purchased, and were often willing to pay more for what was perceived as a premium product.

Price has always been the signifier of quality for many, and wine proved to be no exception to this supposition. Since varietal-labeled wines from Napa were more expensive, ipso facto they were seen as better quality by many, especially wine novices. For the most part, such observations were accurate, as even basic jug wines from California were now well made, balanced, and consistent. Napa had demonstrably taken this to a higher level with their premium varietals and charged several dollars more for their bottles. Since more wineries and winemakers eschewed past practices of bulk production and overplanting to focus on distinctive wines that were distinguishable from other California producers they saw a justification for such pricing. By assiduously embracing years of experimentation and technological advancements, Napa was in the process of creating a domestic market for the production of quality "fine wines." Their timing was auspicious since new consumers were discovering wine, learning to appreciate its finer qualities, and willing to pay more for such an experience.

As noted in chapter one, people have different predispositions toward wine. For some, it is a pleasurable beverage, with or without a meal. For discerning others their simple preference is for the consummate drink, while for others it is a marker of sophistication as a status good. Hedonic factors aside, consumers have shaped the embryonic wine culture inasmuch as they responded to the differentiation that producers were creating. With increased prosperity, added discretionary income provided many consumers with the opportunity to acquire goods that were both self-serving and a visible manifestation of their own accomplishments. As James Lapsley states, "Expensive wine has always been a status symbol, and the growth of connoisseurship allowed Americans to indulge in an aesthetic experience, while simultaneously affirming social status."[12] Not only were Americans drinking more wine that was of overall better quality, they were also willing to spend more to buy premium wine. Many still purchased French wine since it was the known purveyor of quality, but others were becoming familiar with California brands and trying them. By the early 1970s the United States had become the second largest market for fine wines (above three dollars a bottle), and that trend increased by twenty to thirty percent annually during the next decade. Clearly, Americans were becoming more and more aware of wine.

An evolving wine culture was being shaped by corporate and large family-owner wineries through their extensive media campaigns. Those such as Coca Cola's Taylor California Cellars together with large, privately owned ones like Gallo developed sophisticated marketing campaigns to sell their table wines as part of new relaxed lifestyles. Using their advertising skills and large financial resources, they became powerful arbiters in promoting a wine culture among the average American. Alcohol was further stripped of its link to immigrant populations, rosé wines were imagined as the perfect summer drink, and table wines were elevated to a prominent place on the dinner table. Such promotions brought wine to the attention of the general population, and their persistence reinforced the image of the beverage's respectability. As more people became familiar with wine, not surprisingly some developed discerning palates that would lead them to try the premium wine Napa was producing successfully. This trading-up process perfectly suited many Napa producers, who were increasingly moving away from the bulk table-wine market.

While taste and appreciation are difficult to quantify, the increase in wine sales in general and fine wine in particular during the 1970s is indicative of demand-side forces that coalesce with improved wine quality. But they were also powerful agents shaping change among producers. Heretofore, the market for premium wine was circumscribed not just by the ability of producers to deliver a quality product, but also their being able to sell it. That was now changing in ways that benefitted wineries who had invested in delivering a premium product and who were able to meet the vinous needs of a more discerning customer. For consumers, recognition of quality was increasing and Napa brand awareness fitted neatly into this categorization. A seminal event in 1976 would make this reputation building all the more potent.

Reputation Building

Organizations such as the Napa Valley Vintners' Association (NVVA) and the Napa Valley Grape Growers' Association (NVGGA) assiduously worked to both promote and protect the Napa brand since both organizations' inception. As noted in the previous chapter their goal was to present a coherent voice to articulate issues pertinent to quality and integrity, as well as representing winery and grower interests in the broader public sphere. Their efforts had been partially responsible for growing the public recognition of Napa wine. For decades, even as far back as the late nineteenth century, individual winegrowers had been entering their wines in national (and sometimes international competitions), winning medals and gaining accolades for their quality. Napa wineries, as noted in previous chapters, had assumed a dominant presence in such competitions. However, in most cases the wines they were compared to were also from California (or possibly other regions of the United States). Such events marked Napa wines for their quality but did little to demonstrate how they compared with Old World wines that had an established pedigree. All that changed in 1976 when the scion of an English aristocratic family, Steven Spurrier, held a blind tasting in Paris in which the supposedly best of California wines would be compared with those from Burgundy and Bordeaux.

Spurrier had opened a wine shop in Paris in 1970 and then offered wine-appreciation classes in English at the small school that he founded, L'Académie du Vin. His school proved remarkably successful, and in time he attracted French wine experts and sommeliers. In 1975 he decided it would be interesting to compare Bordeaux's first growths in a blind tasting at his school – an intriguing idea that, surprisingly, no one had systematically done before.[13] He felt that such tastings could be good publicity for his shop and wine school.

Since Spurrier's specialty was largely French wine he had never thought much about California wines, and what he knew of the mass-produced ones was notably disdainful. However, several wealthy Americans in Paris who were wine lovers attempted to persuade him otherwise. They applauded the growing quality of wines from California, especially Napa, which led to some positive press coverage. When hearing of this, some California winemakers went to Paris and dropped off bottles of their best wine at Spurrier's shop for him taste. For Spurrier and his shop manager Patricia Gallagher, this was their first experience of California wine that was not designed as a cheap generic table wine. Both were suitably impressed, and thus was born the idea of staging another tasting in which red wines from Bordeaux and whites from Burgundy would be compared with similar wines from California. Wine guide author Robert Finegan (one of the people responsible for alerting Spurrier to the better California wines) produced a list of California wines, all of which were from the newer "boutique" wineries in the state, and none from the old established players. He felt these were most representative of what California had achieved, especially since many used French wines as their benchmark. From this list, Spurrier eventually selected six Cabernet Sauvignons and six Chardonnays from California to compare against four Bordeaux and four Burgundy wines.

For the tasting on the afternoon of May 24 he chose respected French wine experts as the nine judges. He invited press coverage, but was shunned by the French journalists. The only press person to attend was the Paris-based reporter for *Time* George Taber, and he only agreed after persistent cajoling by Gallagher. The judges were told by Spurrier that he had a selection of California wines from small and unknown wineries that he had found interesting, and that he thought the judges might like to try. He added that he was including some well-known similar French wines, and that it would be most objective if the tastings were blind. He asked the judges to provide brief descriptions of the wines' taste profiles and assign a score out of twenty that could be used to rank them. The judges agreed, and everyone sat down to taste from bottles with varietal labels but no indication of their name or country of origin.

The pre-tasting consensus amongst the judges was that distinguishing the French wines would be easy since the presumption of their excellence was without doubt. As they tasted their way through the wines such an opinion was apparently verified as they exclaimed how obvious it was that such and such a wine was French, and likewise how inferior were the apparent California. Ironically, sometimes a top French wine was dismissed by the judges as being weak and insipid since they presumed it was Californian. At the conclusion of the tasting there was widespread disbelief when the results were tallied and it was revealed that the wines with the highest scores in both white and red were in fact from California (a 1973 Chateau Montelena and 1973 Stag's Leap Wine Cellars). These wines had bested a 1973 Meursault Charmes and 1970 Château Mouton Rothschild. Disbelief soon became outrage as the implications of the results were fully realized. The longstanding presumption that Old World wines were naturally superior to New World ones was cast in doubt.

When things had settled down in the hotel were the tastings were held, some of the judges admitted how fine some of the California wines were. Pierre Bréjoux, inspector general of the Appellation d'Origine Contrôlée board, told George Taber after the tasting: "I went to California in July 1974, and I learned a lot – to my surprise. They are certainly among the top wines in the world. But this Stag's Leap has been a secret. I've never heard of it."[14] Similarly, Pierre Tari (owner of third-growth Château Giscours) said: "I was really surprised by the California whites. They are excellent."[15] Others, however, were less gracious, often questioning the veracity of the event and the way it was done.

Not surprisingly, the French press largely ignored the results, and critics soon questioned the viability of the whole process. Admittedly, the methodology was somewhat flawed as six California wines in each category were compared with only four French. Also, wines were given a score out of twenty, and then the final numbers were tallied. Among the reds, Château Haut-Brion received the most first-placed votes, and overall the French wines rated much better than those from California (three of the top four positions). But Stag's Leap won by a mere one-and-a-half points, and this, together with the white results where California was much more dominant, was what would be remembered. The fact that two wines from Napa were deemed of better quality than two iconic French wines, and acknowledged as such by a panel of French wine experts, would not go unnoticed in the United States. From that date onward, Napa was able to gain a heightened level of respectability that none of the earlier domestic US tastings had ever fully delivered.

Chateau Montelena co-owner Jim Barrett was in the south of France at the time of the tasting, and when contacted and asked for a comment on the results he said: "I guess it's time to be humble and pleased, but I'm not stunned. We've known for a long time that we could put our white Burgundy against anybody's in the world and not take a back seat."[16] Stag's Leap's Winiarski was more subdued in his comment when informed of the result. He merely said: "That's nice."[17]

It is interesting to note that Mike Grgich was the man responsible for the Chateau Montelena wine. As noted earlier, both he and Winiarski were recognized for their dedication to excellence and their early realization that quality grapes, when vinified in a technically proficient way, would produce a notable premium wine. Their uncompromising attention to detail often singled them out from others. This was incontrovertible evidence that such efforts had paid off.

Winiarski remained at the helm of Stag's Leap Cellars for many years, enjoying the accolade and reputation that accompanied the tasting. After the tasting, Grgich decided it was time for him to branch out on his own, having tired of working for others for eighteen years. He wanted to buy a small (around two-acre) plot to build a boutique (two thousand-case) winery and tasting room, but ran afoul of the recent agricultural-preserve rules mandating a minimum twenty-acre size. Realizing that if he bought that much land he could not afford to build a winery, in 1976 he entered a partnership with Austin Hills, whose family-owned Hills Vineyard where they grew Chardonnay and Riesling. The deal was a fifty-fifty partnership, and Grgich Hills's winery broke ground in 1977.

Both Grgich and Winiarski epitomized what Napa was capable of. Decades of technical innovation, experimentation, and information sharing among winemakers had created an important pool of tacit knowledge. This community was crucial to the region's evolving reputation because it enabled winemakers to benefit and learn, often from others' mistakes. Innovation often occurs best when formal and informal information is easily transmitted among core practitioners who are able to adopt and modify, but also provide systematic feedback to others. Together with scientific information provided by UC Davies, winemakers had grasped the production techniques that maximized the benefits of varietal planting. But this also alerted them to a better understanding of the unique conditions in the valley, and the extent to which some of this might be responsible for the excellent wines produced.

Terroir, in some implicit way, had always been acknowledged by Napa wineries, but now its articulation was becoming more formative. Old World wines relied on terroir narratives to explain their quality and uniqueness. Napa saw quality more through the lens of science and technology.[18] However, after the Paris Tasting there was renewed interest in the essentialism of place, especially among groups such as the Napa Valley Vintners Association, who for decades had promoted the region's unique aspects.[19] Discussions often focused on the distinguishing characteristics of the region and how this might be formally articulated. Together with the Napa Valley Grape Growers' Association, in 1979 they petitioned the Bureau of Alcohol, Tobacco, and Firearms (BATF) for the introduction of an official viticulture area for Napa. Defining the precise geographical limits for Napa (exactly which areas to exclude) proved difficult, as did the percentage of the grapes that must come from the area to be labeled Napa. Discussions continued for several years until 1981, when the BATF came up with an inclusive definition of Napa that appeared to satisfy most people. Essentially, it stated that Napa County was a geographic area producing quality wine, and for a bottle to be labeled as such it must comprise of at least eighty-five percent grapes from that area. Unlike European appellations that designated quality levels and the types of grapes to be used, the American version of a viticultural district (AVA) merely designated distinct geographical features such as climate and soil that demarcated it from adjacent areas.

Notwithstanding this rather amorphous categorization, the new AVA did lend a coherent identity to Napa – a sense of place from which these remarkable wines were produced. It lacked the regulatory umbrella that delimited Old World winemaking, but that was not surprising given Americans' tentative embrace of government rules. It did, however, prove to be a useful starting point whereby Napa wineries could formally differentiate themselves from other producers in the state, and subsequently use it as a lever to informally develop and enforce quality norms.

Vineyard Management and a New International Partnership

Corporate interest in Napa had continued with further acquisitions during the 1970s. But many corporations realized that the business model of growing grapes and making wine lacked much of the predictability, consistency, and particularly profitability of their other operations. Harvest yields, quality, and grape prices fluctuated widely, often leading to financial losses. Furthermore, many corporate properties were interested in maximum yields rather than focusing on quality and therefore lower yields.

Heublein was one such company that had purchased United Vintners and Beaulieu Vineyards in the late 1960s as part of their diversification strategy, effectively becoming the largest vineyard owner in Napa. In an attempt to gain improved efficiencies by spinning off their vineyard business, in 1973 they restructured the organization, creating a de facto subsidiary called the Vinifera Development Corporation. It was managed by Andrew Beckstoffer, who had helped them run the vineyard management part in its earlier iteration. Essentially, he oversaw the grape-growing side of the new business, and Heublein had the right of first refusal of the grapes, as well as the proceeds.[20] The next few years saw fluctuations in revenue, and eventually Beckstoffer acquired the Vinifera Development Corporation when Heublein finally tired of the vineyard side of the business.

As a manager Beckstoffer become increasingly knowledgeable about grape production and its various limitations. But he was quite prescient in realizing that controlling vineyards that produced high-quality grapes with sufficient operational scale could in the long run be a lucrative business proposition. He also recognized that planting premium varietals such as Cabernet Sauvignon and utilizing technological advances in the vineyard would enable him to get a good balance of quality and quantity. Specifically, he adopted new canopy management practices that adjusted the amount of sunlight that hit the grapes, planted ground-cover grasses to prevent moisture loss, and introduced drip-irrigation systems to maintain adequate soil moisture.[21] He experimented with the planting of multiple clones in specific vineyard sites as part of his strategy to provide vineyard-designated wine to his buyers. Using different clones of the typical blending grapes Merlot, Cabernet Franc, and Petit Verdot together with Cabernet Sauvignon was a way of pursuing diversity and personality for each of his sites.[22] He hoped winemakers would use this to create wines with site-specific typicality, thus further distinguishing their product.

His mission statement was quite precise as well as ambitious when he stated his aim to be, "a large volume, high quality grower of northern California coastal premium wine grapes through the development and application of modern business and viticultural technology, our way to realize above-average returns from farming services and grape sales while building an estate in vineyard real estate."[23] This was the forte of a businessman who recognized the tremendous potential of Napa, but did not want to be a winemaker or own a winery. Many people (correctly) presume that the valued-added comes not from growing and selling gapes but actual winemaking and retailing. Beckstoffer was determined to prove otherwise with premium-quality grapes. Within a few decades, he bought additional vineyards in Napa and adjacent counties and became the largest independent grower on the north coast, with more than 2,100 acres of vineyard.[24] When phylloxera struck again in the 1980s, he replanted his vineyards with genetically superior clones that brought better quality and higher yields. He also replanted with mainly Cabernet Sauvignon, as it was clear to him that this varietal grew exceptionally well in certain parts of Napa and could produce premium-quality wines. His To Kalon vineyard was one such site. It also meant that he could charge a price premium to wineries for the sale of his grapes, and was willing to sign long-term contracts to formalize the arrangement.

A problem that had often plagued growers in the area was the lack of pricing transparency in market activity that had left sellers at the mercy of buyers. In an attempt to rectify this imbalance, the Napa Valley Grape Growers Association (NVGGA), which Beckstoffer helped found in 1975, had actively been lobbying the state for procedural improvements in sales transactions. From its inception, the NVGGA attempted to give a voice to growers who felt that their interests were being overshadowed or even ignored by the owners' organization. One of their early legislative successes resulted in wineries being required to reveal what they paid from grapes. When this information was combined with growers' cost calculations, it became clear that many growers were barely breaking even. Consequently, the NVGGA developed and implemented a formula that established a basic price for a ton of grapes – one hundred times the retail price of a bottle of wine made from such grapes. The influence of growers of premium fruit such as Beckstoffer, whose grapes were in high demand, eventually created some symmetry to what was often a power imbalance between growers and wineries. This reformulated arrangement would persist for the coming decades. Long-term contracts also brought stability to this part of the market by securing sales predictability for growers and guaranteeing a reliable supply of premium grapes for wineries.

New Partnerships

We have noted how partnerships between individuals interested in acquiring a winery provided a way to spread the financial risk and obtain the necessary working capital for operations. And, in the case of Moët-Hennessey's development of Domaine Chandon in Carneros, such foreign investment demonstrated the attractiveness of the valley's potential. But when Robert Mondavi revealed his proposed joint venture with one of Bordeaux's iconic wineries, Château Mouton Rothschild, it surprised many as it seemed a rather presumptuous arrangement that was excessively leveraging the success of the Paris Tasting. It might make perfect sense for Mondavi, but people wondered what was in it for the French.

Mondavi's winery was continuing to expand its operation during the 1970s, with his sons Timothy as director of production and vineyard management and Michael as winemaker and vice president of sales. The focus continued on producing a high-quality wine from what were some prime vineyard sites, including parts of the highly-respected To Kalon. Wine that did not meet the quality specifications for the Mondavi label continued to be sold as bulk wine through the Woodbridge brand. As ever, under Robert Mondavi's tutelage, the winery embraced technical experimentation, searching for better techniques that would improve the quality of the wine, as well as fostering the sharing of operational knowledge in the local community of winemakers and winery owners. Both were seen as crucial for the sector to continue its embrace of quality production. It was well known that Robert was an innovator, especially if it improved the quality of his wine and the reputation of his brand. But he also demonstrated a willingness to work collaboratively. His determination to build a "wine community" in Napa was in fact predicated on the essentialism of cooperation and sharing, even among putative competitors.

Meanwhile, in Bordeaux, Baron Phillippe Rothschild had finally managed to get his Château Mouton Rothschild winery elevated to the status of a first growth. Categorized in the 1855 classification as a second growth, the winery under Rothschild had long thought of itself as really belonging to the top category. He had badgered the authorities for years and earlier struck informal deals with the original four first growths as a way of insinuating himself in their reputational aura.[25] For decades, this bore no fruit, until finally in 1973 the French Ministry of Agriculture bowed to his pressure and gave him the formal prestige that he desired by elevating the château to premier cru status (first growth). This was a remarkable achievement as it remains the only change made to the original 1855 classification. Doubtless, many other lower-tiered wineries would have liked such an accomplishment, but perhaps lacked the influence and persistence of Rothschild.

Secure in his new standing, Rothschild was nonetheless troubled by the stalled economic growth and mediocre performance of the French economy while at the same time he recognized California's increasing ability to make fine wine. He had already started a second-tier wine (Mouton Cadet), into which went a large harvest of grapes that didn't meet the quality specifications for the regular Mouton. This proved to be extremely successful, and in many respects matched what Mondavi had done with his bulk wines sold under the Woodbridge label. Now he thought it might be intriguing to forge a partnership with a California winery to make a New World first-growth wine. He too was not averse to collaboration if he felt it would be advantageous to his overall goals of cementing his reputation and reducing risk.

Following advice from a few friends who were familiar with premium California wineries, Rothschild initiated a number of informal meetings with Robert Mondavi about some form of collaboration between the two wineries. The general idea discussed involved creating a wine made in Napa under the joint auspices of Rothschild and Mondavi and labeled accordingly. Such an idea finally came to fruition in 1978 when Rothschild proposed a formal partnership with Mondavi. The wine would be made from premium Napa grapes using longstanding French techniques and practices. A property for the vineyard and winery was acquired in 1981. The name chosen, after much discussion, was Opus One, and to symbolize unity both men had their profiles artistically rendered on the label. Their goal was to produce a limited-production Bordeaux-style red wine, aged in French oak barrels and sold at a price premium to indicate its quality and desired status.

The first vintage announced was 1979. The wine was made jointly by Rothschild's experienced winemaker, Lucien Sionneau and Mondavi's Timothy Mondavi. Early on, however, the two differed as to whether the drink would be a food-friendly elegant wine in the French style or a bold, intense one like that which Napa was beginning to produce.[26] Even by the late 1970s, the stylistic character of premium wines from Napa had become quite distinctive, and this distinguished them from Old World wines. The eventual compromise was a combination of the two styles. The 1979 and 1980 vintages were combined for the first release, and the wine was priced at fifty dollars a bottle (four to five times the price of the average Napa bottle). In spite of the high price, most of the press coverage was positive, and it was abundantly clear that the partnership endowed Mondavi with a heightened status as well as self-confidence about what he was doing. For Rothschild, it gave him better access to the US market for his château wines, as well as spread the risk for his business in the uncertain economic times in France. Crucially, both men shared the same opinion about selling premium wine through the image of a luxurious lifestyle. They envisaged wine as the consummate experiential good for those who could afford its expense, an indication of status and sophistication. It would be seen as a lifestyle marker to differentiate connoisseurs from the everyday consumer. Expensive wine was a status symbol, and Napa producers were beginning to realize that the region's nascent brand conferred the ability to charge a price premium. In fact, since the Paris Tasting, many Napa producers had been able to raise the price of their wine as demand increased. Opus One's pricing strategy was thus a logical continuation of this development. It was, as Robert Mondavi himself claimed, the first "ultra-premium" wine.[27]

By leveraging the association with the first-growth classification of Château Mouton Rothschild, Mondavi was able to elevate the reputation of his own winery. Everyone knew that Opus One was a joint venture. This was a de facto anointing of Old World traditions and reputation for a New World winery. The lavish winery that was constructed in Oakville bore testament to the "no expense spared" approach by the partnership. The winemaking followed many of the traditional Bordeaux practices, and the finished wine was intended to be comparable to a first growth. Mainly Cabernet Sauvignon blended with small amounts of Cabernet Franc and Merlot (a similar formula to that of Left Bank wines from Bordeaux), the resulting wine was an interesting combination of American commitment to technology (much testing and analysis) and the more intuitive French art of blending (*assemblage*). Like many classic French wines, however, it was tannic and needed years to fully mature, hence some of the first tastings produced skeptical comments from the wine press about the quality. But the first vintages sold out upon release, so perhaps the hype was sufficient to dull any critical concerns over initial tastes.

After a steady growth rate from the late 1960s, the number of wineries in Napa went from 51 to 110 between 1977 and 1981. This significant increase can be partly explained by the Paris-tasting publicity. Some of the new winery owners were decried as moneyed entrepreneurs who brought their fortunes to buy into a lifestyle, attracted by the charm of the valley and the status of owning a winery. But there was nothing new in this, since others had founded wineries from fortunes made elsewhere from the late nineteenth century. Other newcomers had less-impressive financial resources, only a determination to make a quality wine and be profitable. Cain, Groth, Honig, Chateau Boswell, Spotteswood, Dunn, and Shafer are examples of such new wineries founded in the late 1970s and early 1980s. They would become established brands in the decades to come and set the stage for patterns of subsequent faster growth.

Conclusion

It is easy to lapse into great-person narratives when explaining the 1970s growth in Napa. The likes of seminal figures such as Robert Mondavi, Andrew Beckstoffer, Mike Grgich, and Warren Winiarski are such an integral part of that decade's changes that it is easy to forget others who quietly went about their commitment to making premium wine. The Paris Tasting threw the spotlight on certain wineries, but others whose products were not entered into the competition were also making fine wine. Some of them were new wineries, others new owners who had renovated an existing site and brought in a new winemaker. In most cases, they adopted practices that were quite different from those of their predecessors. Whereas in the late nineteenth century innovations were often associated with new grape varietals, in the twentieth they typically involved new techniques and technology.

The 1970s saw both the continuation of a gradual accumulation of knowledge and resources that were essential to the making of a quality wine as well the introduction of distinctive new practices. Technical experimentation, improved winemaking skills, better-quality grapes and understanding the best vineyard sites combined in new operational paradigms. Bulk wines and blends made from grapes such as Alicante Bouschet and Petite Sirah were replaced with lower-yield but higher-quality Cabernet Sauvignon and Chardonnay. Even Zinfandel was proving less desirable to plant – not because of quality concerns, but largely because the finished wine could not command the price premiums of Cabernet.

The reality was that growers as well as wineries had managed to solve financial problems by planting and vinifying higher-quality grapes that commanded a higher price. This was all part of a trend in which the market for Napa wines was increasingly seen in terms of premium and even ultra-premium products designed for a more discerning and affluent consumer. Wine consumption continued its increase, and within that market there was scope for product differentiation that Napa was able to fill. Bulk wines were commodity products that were largely consigned to the past or remained the workhorse of a few larger wineries that used such operations to cross-subsidize their premium products.

As wineries were developing their productive capabilities, taking advantage of opportunities that came from knowledge dissemination, some were able to differentiate their products, albeit often subtly. Identifying key vineyard areas that would enhance their best practices provided, when combined with incipient brand building, more distinctive reputations. Such innovation enabled winegrowers to recharacterize the market for their wine on terms suited to their emerging capabilities. In this sense, they were able to shape the market, whereas heretofore the market had generally shaped them.

The enterprise-growth dynamic that we have seen in this period was also marked by continued institutional changes. Universities gained further legitimacy with the development of professionalized viticulture and oenology programs which stimulated formal knowledge transfer and created systematic credentials to underscore occupational professionalism. Legislation creating the Agricultural Preserve limited suburban development but also pushed up the price of potential vineyards because of minimum size restrictions. Starting a winery had become more capital intensive, and with fixed-asset costs rising, small-scale quality production meant marginal costs remained high. Newcomers needed progressively greater financial resources to become viable (and hopefully profitable). Finally, organizations representing growers and winery owners (plus informal ones among winemakers) had been able to coordinate promotional activities as well as lend legitimacy and coherence that helped frame an evolving normative operational framework. Cooperation between winemakers had been paramount in the development of informal benchmarking practices, and this was given saliency when the Paris-tasting results were announced. The two winemakers responsible for the winning wines had continued the practices of André Tchelistcheff in his unwavering commitment to excellence, but had the resources that he so often lacked. The continued willingness of many parties to share information proved crucial for the sustained-growth dynamics of the sector. It mitigated the need for excessive trial and error and more clearly defined best practices, and enabled newcomers to participate in the ongoing capability development. And when combined with an external market in which demand was becoming differentiated, plus an institutional framework that augmented growth, wineries were well positioned to take advantage of what were high value-added product propositions.

Notes

[1] Grgich was responsible for all aspects of winemaking at Chateau Montelena, from the vineyard to the cellar, monitoring how the grapes were grown and the quantities harvested (he preferred small clusters for heightened quality), and often ignored the advice from university experts on sulfur application to prevent mold and mildew by arguing that nature needs to be left alone to work its magic on the grapes. His entire winemaking process was described by George Taber as a gentle, reasoned determination.

See George Taber, *The Judgment of Paris* (New York: Simon and Schuster, 2006), 143–5.
[2] "1980 Market: 490 Million Gallons," *Wines and Vines* (September 1972), 23.
[3] James Lapsley, *Bottled Poetry* (Berkeley: University of California Press, 1996), 202.
[4] Founded by intrepid pioneer Josephine Tychson in the 1880s, the first woman to own and operate a wine estate in California, it was named Tychson Cellars. After phylloxera struck it was sold to Italian immigrant Antonio Forni in 1894, who renamed it Lombarda Cellars. Despite being able to make and sell sacramental wine during Prohibition, he was eventually forced to sell, and the winery was acquired in 1939 by three Southern California property developers: Charles Freeman, Mark Foster, and Abbey Ahern. The name Freemark Abbey came from an abbreviation of each of the partners' names. When Ahern died in 1959, the winery closed until its 1967 purchase by the Carpy partnership.
[5] Quoted in Lapsley, *Bottled Poetry*, 202.
[6] Ibid., 203.
[7] The 1973 vintage was large, as was the 1974. Quality was excellent, but so was quantity, and this led to a price depression.
[8] Napa County Agricultural Commissioner, *Annual Report*, 1967 and 1980.
[9] Charles Sullivan, *Napa Wine* (San Francisco: Wine Appreciation Guild, 2008), 299.
[10] Lapsley, *Bottled Poetry*, 200.
[11] Ibid., 198.
[12] Ibid., 199.
[13] Taber, *The Judgment of Paris*.
[14] Ibid., 205.
[15] Ibid., 205.
[16] Ibid., 207.
[17] Ibid., 209
[18] For a fuller discussion of this often-contrasting stance see Ian M. Taplin, "Narratives of Science and Culture in Wine Making," in *Routledge Handbook of Wine and Cult*ure, eds. Steve Charters, Marion Demossier, J. Dutton, G. Harding, Jennifer S. Maguire, Denton Marks, and Tim Unwin (London: Routledge/Taylor Francis, 2021).
[19] Lapsley, *Bottled Poetry*, 205.
[20] W. A. Beckstoffer, "Premium California Vineyardist, Entrepreneur, 1960s to 2000s," Regional Oral History Office, Bancroft Library, Berkeley, University of California, 120–1, https://www.lib.berkeley.edu/libraries/bancroft-library.
[21] John Briscoe, *Crush: The Triumph of California Wine* (Reno: University of Nevada Press, 2018), 268.
[22] Ibid., 269.
[23] Ibid., 266.
[24] Ibid., 266.
[25] In the 1920s, Rothschild instigated the Association of Five where Mouton-Rothschild joined with his neighboring first-growth wineries Haut Brion, Latour, Margaux, and Lafitte. He encouraged them to develop chateau bottling and promote

their wines as glamorous products. This was a considerable achievement for what were publicity-shy owners. See Julia F. Siler, *The House of Mondavi: The Rise and Fall of an American Wine Dynasty* (New York: Gotham Books, 2007), 161–2 for a fuller discussion of these earlier efforts.

[26] Ibid., 172.

[27] Robert Mondavi, *Harvests of Joy: My Passion for Excellence* (New York: Harcourt Brace, 1998), 221–2.

# CHAPTER FIVE

NOW IS THE TIME FOR FINE WINE

During the decades following the Repeal of Prohibition, Napa gradually assumed the identity of not just a wine-producing region but one where premium-quality wines were more consistently being made. Its putative reputation had taken shape during the 1960s and acquired legitimacy following the Paris Tasting in 1976. Even the most ardent critics, such as Old World wine connoisseurs, had to admit that something good was being produced in Napa – wines that could stand with the best produced in France. Decades of technological innovation, experimentation, and the arrival of newcomers with passion, energy, skill, and financial resources resulted in the production of more consistent high-quality wines. Furthermore, a culture of cooperation and technical information sharing among winemakers was instrumental in generating the organizational framework that bestowed operational efficiency. Knowing what grapes worked best on what terrain and how to balance yields with quality was shared informally. Best practices became formalized as winemakers learned from others and were able to implement accordingly, especially since many now had more plentiful resources. Such activities enabled wineries to shape the market, differentiating their product from table wines and allowing them to continue to charge a price premium.

By the 1980s, American consumers were also demonstrating a growing enthusiasm for domestic wine. Consumption was increasing, and not only were people drinking more wine, they were drinking better-quality wine. While not necessarily euphoric in their changed behaviors, they nonetheless revealed an enhanced discernment of premium varietals. Trading up in their consumption suggests both improved material conditions as well as an appreciation for wine as something other than a mere commodity. Wine figured more prominently in social life and around food. Consumers appeared to be more confident in buying wine and often used it as an indication of social identity.[1] The increased demand for premium wine was a boon for Napa producers whose product naturally fit this category. It also proved a stimulus for new winery owners who were attracted to the industry for its lifestyle and a desire to excel at making a product that evoked sophistication and status – precisely what Napa wines were acknowledged to embody and what consumers wanted.

The next two decades would be crucial for Napa as it cemented its reputation by forging a more pronounced stylistic identity, attracting industry entrants, both private and corporate, with extensive resources, and negotiating a sometimes-tortuous path with replanting to circumvent further phylloxera outbreaks. Winemakers increasingly understood the character of the grapes and how they were expressed in the finished wine. For that reason, Cabernet Sauvignon and Chardonnay had become the preferred varietals since they proffered distinctive styles and taste profiles that reflected the growing conditions in Napa. But while science had enabled winemakers to make a consistently better wine, it didn't necessarily tell them why certain sites produced wines with particular flavor profiles and how those differences could be distinguished. The differences might be subtle but were increasingly recognized as a product of topography and different vineyard sites. The American enthusiasm for technological solutions permeated most vineyard and winemaking activities, but might the clues to stylistic difference also lie in the soil? Perhaps there was something to the notion of terroir after all, especially if stripped of its sociocultural baggage. If this is the case, how could one better understand and categorize these different characterizations?

This would prove to be an interesting digression for an industry whose success was the pursuit of a formulaic agenda where science and technology ruled supreme. Now it was moving into unchartered territory, or at least considering narratives that transcended a more manipulative winemaking approach. Subsequent acknowledgment of site specificity enabled wineries to engineer the true expression of their locale, thus providing another variable in the search for perfection.

Concurrent with these changes, scientific resources continued to be formally disseminated from the universities, with formal training programs complementing hands-on experience for winemakers. The profession became increasingly credentialized, and clearly the rise in consistent-quality wine was testament to the success of this institutional support. But, for some, the 1980s also epitomized the tyranny of technology when university recommendations on disease mitigation proved less auspicious. This was particularly pertinent when university specialists argued that old St George rootstocks should be replaced by a supposedly phylloxera-resistant new rootstock (AxR#1). When it became patently clear that this new rootstock was not disease resistant it forced those who replanted with it to replace it again when their vines succumbed to the disease (I will discuss this issue later in the chapter). Fortunately, by the late 1980s wineries were in a better financial position to rip out and replant vineyards, and they invariably did so with premium varietals such as Cabernet Sauvignon. However, the whole process raised concerns about the integrity and reliability of some academic research.

Having mastered techniques to improve quality and manage vineyard operations, wineries realized that the path to excellence came not just from finding the right places to grow but which were the best grapes to make the finest wine. By the 1980s it was patently clear that Cabernet Sauvignon, Chardonnay, and Pinot Noir provided the best-value proposition for an area where land prices were rising and acreage increasingly scarce. Traditional blending varietals such as Zinfandel simply could not command the high prices that the other three delivered. Despite a brief period of acclaim following the success of White Zinfandel, Red Zinfandel would become more of a niche player in subsequent decades, migrating west to adjacent Sonoma County where it has nonetheless remained a signature grape. Land in this county is more plentiful and less expensive than Napa.

As consumer tastes changed and Napa demonstrated its ability to make premium wine from the three aforementioned grapes to satisfy this evolving palate, perhaps not surprisingly more vineyards were replanted with them. Napa Cabernet Sauvignon not only grew well but was increasingly recognized for its muscular and robust style which appeared to resonate with customers and critics. Arguably, this style was the result of geographic and climatic specificity, but also reflected when the grapes were picked and how the wine was made. The wines were richer and riper tasting, more opulent, and often fruit forward. This distinctiveness served to differentiate Napa Cabernets from others in California, and in the years that followed came to be the acknowledged characteristic of the region's primary grape. But not everyone agreed with this, and for some the wine was seen as a product of over-manipulation – an obsessive technological rigor constraining a more-natural expression. This debate surfaced in the 1980s and continues to this day, albeit mediated via the role of influential critics, whom many say favor Napa's overripe style. The role of critics became a powerful arbiter of Napa's continued rise, and is discussed later in this chapter.

It would be churlish to not mention trends at the opposite end of the value spectrum – the success of the so-called super-value wines. While most Napa producers unambiguously bought into the premiumization as the preferred route to success, there were some who realized the marketing potential of selling a mass-produced table wine at very low prices that was labeled Napa. Even if the wine wasn't from Napa grapes (often it wasn't), as long as it was bottled in Napa, the Napa label appeared to be a legitimate designation. Thus was born the "two buck chuck" phenomenon, when Frank Franzia muscled his way into the valley and provoked an outcry over what actually could constitute a Napa-branded wine. As we shall see later in this chapter, for a brief period his foray entertained value-oriented consumers and made much money for himself, as well as the attorneys from both sides in the inevitable legal proceedings that ensued.

What Consumers Want

Throughout this book I have argued that market forces shape how firms and clusters evolve. Notwithstanding supply-side forces of eager entrepreneurs, certain aspects of why products succeed are determined by what consumers are willing and able to buy. Demand forces can be powerful stimulants for production, especially if firms have requisite resources and capabilities to meet such demand. In the case of the American wine consumer, tastes were changing as people became more affluent and better informed about wine. A wine culture was emerging, and wine was no longer seen as an immigrant beverage or even more negatively as a drink of choice on skid row. Many consumers were now more confident in their buying decisions, seeking out distinctive flavor profiles and styles of wine that fit different occasions. Per-capita wine consumption increased during the early 1980s, and a segment of the population was willing to pay a price premium for fine wines.[2] The French had historically cornered that market, but increasingly Napa was coming into play, especially with Cabernet Sauvignon and Chardonnay. Such growing discernment meant consumers were willing to buy expensive domestic wines, providing they were of high quality and signified the status attributes that accompanied such a purchase.

As wine became a social beverage, inevitably it also became part of what sociologist Cecilia Ridgeway refers to as a cultural schema organizing many of our interactions.[3] Living amid hierarchies that we implicitly create and nurture as part of our identity formation in routine interactions, we often value and pursue status. What we buy is part of this process since it signals to others our desired position. Wine plays an important role here because it connotes sophistication and thus allows a person to inform others of their preferred identity. The more expensive the wine, the higher the presumed/desired status. As noted earlier, Americans were not just consuming more wine than in previous decades, they were also buying premium wine. More than any other region in California, Napa was able to capitalize on this trend.

Napa's emerging brand distinctiveness was the product or several forces, both historical and recent. For a start, most Napa wineries had by now embraced an American vision of winemaking, after tweaking what the Old World had to offer and immersing it in scientific expertise. Such a vision, as Matt Kramer succinctly states, is "quantitative, methodical, verifiable."[4] It relies on the adoption of proven techniques and a distinctive approach that considers climatic details (degree days), soil, trellis spacing, vinification methods, and a plethora of incremental improvements in the cellar and fermentation procedures. But most of all, these are seen as ways to develop a more rigorous technically mediated approach that systematized winemaking and guaranteed a more predictable outcome. Such operational improvements enhanced consistency and allowed winemakers to more proficiently deal with harvest variations. This was noticeable in the mid to late 1980s when a series of excellent vintages further cemented the valley's reputation. The emphasis on high quality is also testament to the increased focus on top varietals. Vineyard acreage did not increase, but the percentage devoted to Cabernet Sauvignon and Chardonnay did.[5] And despite a strong dollar that made Bordeaux wines more affordable, Americans remained enthusiastic consumers of high-quality Napa wines.

Ownership transactions continued apace during the 1980s with two trends. First, and most significantly, entrepreneurial interest among resource-rich as well as experienced wine-industry specialists brought new faces and a further quality commitment to the valley. Second, particularly in the latter part of the decade, both domestic and international corporate interests saw the takeover of some of the historic properties as well as wineries that had been founded in the past twenty years. These trends affirm the growing dynamism of the sector and indicate the attractiveness of the industry as a business proposition. In many cases they furthered innovation and provided enhanced operational rationalization. The robust social practices that facilitated learning and cooperation were all constituent parts of the local socioeconomic forces that shaped production and performance in Napa.[6] They built on conventions as well as modified certain practices such as varietal selection where the emphasis on premium grapes continued. This enhanced the competitive viability of the sector by integrating newcomers (and their ideas) with incumbents, and forged a community of owners who embraced a fairly unified vision for the industry's growth trajectory.

Increasing wine consumption and capital availability are integral to the opportunity-space development that is crucial in niche-market growth. After the end of Prohibition, Napa witnessed what we might refer to as generalist firms (combining bulk-commodity production alongside premium varietals) and smaller specialist firms with smaller production volume and a focus on high-quality wines. The extent to which these two types compete in the same market has been discussed extensively elsewhere.[7] They both brought capabilities and resources that, as noted above, endowed the region with its distinctive character and reputation. But as the density of wineries increased in Napa, it attracted new foundings by invariably wealthy individuals who leveraged collective-organizational knowledge and capitalized on institutional support, but also brought new capabilities that furthered operational upgrading. Such entrepreneurs replicated the efficiencies of incumbent wineries and added further specialized skills and a heightened commitment to making excellent wines. They were specialist firms whose presence did not necessarily crowd organization density. Instead, their complementarity proved valuable for further reputation building by contributing significantly to the brand narrative. This is consistent with what organizational scholars have argued when specialist organizations proliferate as industries mature, eventually endowing what had been a niche sector with enhanced dynamic capabilities.[8]

A Few of the New Owners

While each person who came to Napa to start a winery had his or her own rationale for the move, the following three individuals are somewhat representative of the trends in terms of motivation and presumably their eventual success. Did every new winery owner succeed as these did? Probably not. But most came with a similar desire and passion that led to different trajectories. These three individuals came in the late 1970s and early 1980s and have left a mark on the industry. They are emblematic of the new ownership trends, and for that reason it is worth narrating their stories. Others are referred to briefly in the subsequent discussion. This is not intended to diminish the credibility or success of the latter, as in fact many established what have become iconic wineries, and in some case their stories have been told elsewhere.[9]

Shafer Vineyards

What compelled a middle-aged senior manager in a textbook company, living in Chicago with his family, to uproot everybody in 1972 and move to an old property on the eastern side of the Silverado Trail that he had bought? This is the story of John Shafer when he decided to start a new career and chose wine as the desired option. In 1972 he flew to California to look for property on which he could start a winery and was directed to Napa where, with the help of a real-estate agent, eventually found 209 acres on a rugged hillside in the southeastern part of the valley. He had become fascinated with wine, but largely as a businessmen who recognized the increased American interest in wine, higher disposable incomes, and the fact that Napa appeared to be the center of new developments in the industry. He liked the idea of working outdoors, and the more he read about the wine industry the more he seemed committed to finding a way to become part of it. He knew that hillsides were drier than the valley floor and appeared to do well with the terracing that southern European vineyards had adopted. But perhaps most of all he was looking for something new to do – a career change, and a challenge.

After planting vines and growing grapes that were sold to others, Shafer broke ground on a new winery in 1979 with a focus on Cabernet Sauvignon. There was still uncertainty that this would become the signature grape at that time, and white grapes appeared to offer more immediate financial benefits. But by then Shafer had dreamed of making a world-class Cabernet from his hillside grapes – this was the varietal he believed best suited to the terrain, and he was determined to achieve this. The first years were difficult financially, and there was always a "hands-on family approach" to most day-to-day activities. John Shafer was resourceful, but he did not have deep pockets like some other newcomers.

His son Doug became winemaker in 1983 after a brief career as a math teacher following his graduation from the oenology and viticulture program at UC Davis. He was joined in 1984 by Elias Fernandez as assistant winemaker, the latter becoming head winemaker in 1994 when Doug assumed the role of company president. Both learned from hands-on experience as well as participating in the monthly meetings of the Napa Valley Technical Group that had been founded in the 1940s by André Tchelistcheff. Including listening to formal presentations by UC Davis researchers, the group exchanged valuable information about their own practices and problems. The importance of this tacit knowledge cannot be understated as it helped many newcomers to grasp routine issues that emerged on their properties. As Shafer noted, he and Elias didn't miss a meeting for years as "it was a vital source of up-to-the-minute data, approaches and opinions reading the things we were often in the middle of trying to understand and resolve."[10] During this time, the winery employed a consultant winemaker, Tony Soter, who had worked at some of the other up-and-coming wineries in the area. He showed them how to resolve some problems (including how to mitigate excess sulfides), and, during his two-year stint with them, helped them better understand the many nuances of their vineyard site.

Eventually, Shafer went on to produce a range of distinctive, full-bodied wines that continue to receive critical acclaim from the top experts. Such endorsements have been beneficial to the winery, but are also a testament to the embrace of a style and profile that have come to be quintessentially Napa – whether the full-bodied reds or the distinctly oaky and buttery Chardonnays that many American consumers deem their preferred choice. In other words, the winery's evolution focused on making something that consumers like, and its being able to make it very well and in significant quantities to be profitable. As Shafer notes, "we keep producing wines that are true to the area. They are big, bold luscious reds that fortunately critics like Robert Parker really like."[11] He also recognizes the challenge of keeping the brand fresh and he still tries to keep the focus on four wines rather than produce a wide range, as others of their production size do. Currently at approximately thirty-two thousand cases a year, they have certain economies of scale but are clearly neither a "boutique winery" nor a mass-producing one. In this respect, they resemble a château production model – a size and scale that permit efficiencies without losing sight of the meticulous attention to detail necessary for an ultra-premium product.

In discussing his wine and the winery, Doug can be quite expansive but also pragmatic. He is emphatic about how they remain a family winery and the informality that bestows upon so many of their relationships with the public. But given the demand by the public to taste his wines, he recognizes that there will be limitations since their tasting room is small and by appointment only. He wants visitors to have a memorable experience, trying the various wines while gazing out over the estate southward to bay, but only a handful of visitors ever get to experience this.

Dominus

It is difficult to ascertain definitively that Pomerol's Château Pétrus general manager Christian Moueix entered into a partnership with the late John Daniel's daughters to acquire Napanook vineyard in 1982 because he was inspired by the success of Napa wines at the Paris Tasting. But given his enthusiasm for embarking on a venture thousands of miles away from where his iconic winery was located does suggest more than a simple leap of faith.

He clearly recognized the potential of Napa as a place where great wines could be made, especially if one brought the requisite skillsets and viticultural experience to complement the extant practices in the valley. When he eventually bought out the daughters and built a winery, he brought over his French oenologist Jean-Claude Berrouet to make the wine. He never claimed he was trying to replicate his Old World Pomerol, and in fact used seventy-five percent Cabernet Sauvignon as the major varietal in his wine (alongside some Merlot, Cabernet Franc, and Petite Verdot) rather than mainly Merlot, which would have been the case back in France.[12] Perhaps this was because the soil was completely different in Napa from the heavy clay of Pomerol, and if anything, as a Frenchman he did more than pay lip service to terroir. If he was to make an excellent wine in Napa then he needed to use the grape varietal that apparently grew best there. This was Cabernet Sauvignon. His first vintages (1983 and 1984) were released in 1988 with a price of forty dollars – considerably cheaper than Pétrus, but still a high price for a Napa wine with no track record. The initial critical reviews of his wine were positive,[13] and he continued to invest in making a wine that was stylistically somewhere between Old and New World.

Harlan Estate

Close to the Martha's Vineyard site near Oakville, real-estate developer and co-owner of Meadowood Resort Bill Harlan bought vineyard land in 1984 to start his own winery. He had been looking for land in Napa on and off since the 1970s, and when he acquired Meadowood his intention was to transform the property's golf course into a vineyard. But the soil wasn't good enough and the climate too cool, and Robert Mondavi managed to disabuse him of this idea. So, together with other partners, he acquired the old Sunny St Helena winery,[14] renovating and renaming it Merryvale Winery in 1991. Working there with his winemaker Bob Levy, buying grapes from independent growers, he wanted to learn where the best grapes were grown. Merryvale was an empirical learning experience with some trial and error, but also grounded in a rigorous attempt at discernment. When he finally bought the land for Harlan Estate his vision was quite simple – be meticulous in attention to detail, hire skilled personnel, and have a long-term vision for the winery (his two hundred year plan). From the beginning, he was quite explicit in stating his goal to create a first-growth wine in Napa.

The land he chose was east-facing forested hillside on the western side of the valley, and he spent much time and money clearing it. His first plantings succumbed to phylloxera, so he started again. When he developed Harlan Estate he decided to focus on Bordeaux varietals with predominantly

Cabernet Sauvignon, as he believed this varietal would be the quintessential expression of Napa. He deemed his early vintages in the late 1980s as lacking the requisite quality he sought. The year 1990 saw his first vintage released, but it did not go to market until 1996. Harlan wasn't completely certain of the quality, so decided to wait and see how it aged.

In buying grapes for the Merryvale Winery he had experimented with multiple growers in numerous locations in an attempt to discern the best sites. Specifically, he wanted to determine how much difference there was between sites to understand not just the character but the nuances between them.[15] Once he had a better idea he selected five of them (six to twelve-acre vineyards), from which he sourced grapes through long-term contracts. Thus, the single-vineyard bottlings for his next label Bond were created. This was a deliberate attempt to make very specific estate-based wines where the differences were subtle but noticeable to those who were discerning customers. He also created a second label, The Maiden, from grapes he felt didn't quite meet the rigor for his Harlan Estate wines. Recently, he developed another property close to Harlan Estate called Promontory, from which he makes a wine of that same name.

He has always claimed that he wants to make wine that reflects the mood of the land – a unique expression of place. He is not a terroirist in the French vein since he has fully embraced the technical paradigm of winemaking, but in recognizing unique properties that can benefit from an almost Baconian operational rigor he has been able to make wines that critics and consumers recognize as excellent. When released in 1990, his wines were priced at sixty-five dollars a bottle, which he felt was comparable to similar ultra-premium wines from the area. As he has been quoted as saying on this pricing strategy, "Part of our job is to let the world know that Napa wines are at the same level of price and quality as the best of Bordeaux."[16] This deliberate positioning is important for the evolving reputation building of not just his wines but others in Napa who share the same vision of excellence. Finally, as Napa wine-critic James Laube noted about the overall presentation of his wine, "Branding is impeccable. Packaging is precise. No detail is overlooked. The labels themselves are representative of Harlan's pursuit of authenticity and historicity, made by an engraving company that until the mid-19th century designed US currency notes."[17] Several others would follow this approach in the next decade with small-production, limited-availability, and high-priced Bordeaux blends.

Like some of the other winegrowers who started around this time, he has been unequivocal in his commitment to make the finest wine possible from an area that he felt could consistently deliver it. He was fortunate in being able to marshal extensive resources that allowed him to postpone profitability for the first twenty years, and it wasn't until 2004 that the cash flow was positive. As his estate director Don Weaver was quoted as saying, "if he just wanted to make money, he would have stayed in real estate."[18] In his long-term vision he was not different from others who similarly brought resources from other ventures to enable them to focus on rationalizing operations in the pursuit of quality. This combination of vision and capital is important since, as we noted earlier, John Daniel at Inglenook was committed to and succeeded in making exceptional wines, but eventually lacked the resources to continue in that quest.[19]

Other notable entrants to the industry included businessmen Joseph Phelps, who in 1973 bought hundreds of acres on the northeast side of the valley off the Silverado Trail. Initial plantings included Riesling and Sirah, but it was the Bordeaux blend with the proprietary label Insignia, released in 1974, that brought recognition to the winery. The wine was made from the finest lots of the vintage with Cabernet Sauvignon the primary grape. According to winemaker Craig Williams, the wine was a true expression of place (Napa) and style (Bordeaux blend).[20] When Williams took over production at the winery in 1980, he was another beneficiary of the Wine Technical Group's informal wisdom. According to his recollections, each person paid annual dues of twenty dollars and they met monthly at the Silverado Tavern to figure out how to deal with many of the small problems they faced on a daily basis. He recalled one such evening when they invited a speaker to explain SO_2 problems during yeast fermentation: "None of us had quite figured out the problem and this proved amazingly insightful. Afterwards we all tried the wines that each of us had brought and offered critiques. We needed to share and communicate. It was a true community of learning. It proved to me yet again that the wine business was born out of mitigating flaws in wine making."[21] Again, the story is one of collective and cooperative endeavors, plus a healthy dose of self-criticism and a willingness to standardize practices to obviate idiosyncratic occurrences.

Further south on the Silverado Trail is Silverado Vineyards, founded in 1981 by Disney heiress Dianne Miller and her husband Ron. Initially, they sold their grapes to surrounding wineries, but eventually built a winery and added to their vineyard acquisition, which currently has almost six hundred acres under vines from which they produce a wide range of wines. Unlike some other more-specialized wineries, they produced a wide range of varietals, but in each case the aim is to offer a quality wine at different price points. By seeking a broader customer base they were tapping into an increasingly discerning consumer who viewed consumption as part of a variegated lifestyle – different wines for different occasions.

At the opposite end of the spectrum there were individuals who were more modest in their land acquisition but not necessarily in their goals toward making a perfect wine. For example, Richard and Ann Grace planted one acre of Cabernet Sauvignon in 1976 and had their first vintage in 1978. An additional acre planted in 1986 succumbed to phylloxera and had to be replanted. A third acre was added later. Production was and is modest, but the goal was to provide small quantities of a finely crafted wine. Theirs is the quintessential boutique winery.

On the eastern side of the valley, up on Pritchard Hill, businessman Don Bryant acquired a thirteen-acre estate in 1986. With wealth from his various entrepreneurial activities he was able to focus on producing small quantities of the best-quality grapes in a finely craft Cabernet Sauvignon that continues to win awards and is much sought after, albeit sold on allocation. Further north on Howell Mountain, also on hillside slopes, trained winemaker and self-described farmer Randy Dunn planted fourteen acres of his own vineyard and released his first Cabernet Sauvignon in 1979. While starting his own winery, Randy worked at Caymus from the mid-1970s to early 1980s gaining hands-on experience while accumulating the capital he needed for his own winery. Eventually getting his winery bonded in 1982, it remains an extremely modest operation, but his focus upon making wine with modest alcohol levels that is designed to be aged rather than drunk young places him in opposition to the fashionable trends in Napa towards over ripeness. His wines continue to be much sought after yet the property lacks any of the pretentiousness that some Napa estates have developed. Like Bryant, the wine is sold on allocation but his background and resources were less financial than human capital. At this time, it was still possible to buy land on hillslopes for not outrageous sums so someone such as Dunn could just about manage it.

Also on the lower slopes of Howell Mountain is Viader, created by Delia Viader in 1986. From an academic background and with the financial support of her father, she cleared the steep hillside of shrub in what was an expensive endeavor that would probably run afoul of environmental regulations today. Her 30 acre site produces low quantities of fruit from a high density vineyard, mainly Cabernet Sauvignon and Cabernet Franc. Her belief in the latter stemmed from a desire to make a balanced wine similar to a Right Bank Bordeaux although the wine itself is quite robust. Although lacking in a wine-industry background, she is quick to note how the community of winemakers and owners in the area were a valuable source of support in her initial years. Whether it was Robert Mondavi loaning her equipment for the weekend and for generally being an inspiration for what she went on to accomplish to others who offer advice about all aspects of winemaking, the ensuing support system was, according to her, tremendous.[22]

Finally some grape growers continued to be actors in land transactions in the valley during this time. Most notably was Andy Beckstoffer's 1990 purchase of the 225-acre Beaulieu Vineyard #3 – one of the great Cabernet Sauvignon producers of previous years which was seen as a cornerstone for his continued emphasis upon acquiring prime vineyard land for premium varietals. He strongly believed that the valley's future lie in the production of low yield, high quality varietals such as Cabernet Sauvignon. More than other grapes this would enable growers (and wineries) to maximize the value-added potential and offset the heightened production costs. A renowned attention to detail and rigorous viticulture standards enabled him to charge a price premium for his harvested grapes. And his vineyards were increasingly recognized as the source of some of the best fruit in the valley.

These are just a few of the stories behind some of the many Napa brands that we are familiar with today. Together they epitomize the ownership changes that were occurring from which the region's identity was being forged. Each of the first three discussed has similarities with many of the others that have been founded during this time period. This includes individuals who were seeking a new life and saw wine as an inspiration for their dreams; those who had extensive wine experience but recognized the potential of Napa as an area akin to Bordeaux; and those whose extensive financial resource allowed them to indulge their wish to create something remarkable and long lasting. Precisely because of their wealth, some were able to focus unambiguously upon the pursuit of excellence without immediate concerns for positive revenue streams. For others with more modest means, their ambitions in terms of output were more circumscribed even though they shared the commitment to quality that would increasingly define wines from the valley (and the price points that they were able to command). To varying degrees they all recognized the inherent value of collective organizational learning and possessed a reasonably unified vision about the valley's future direction. They saw Napa's success as a communal endeavor, premised on key varietals that would deliver price-premium wines catering to discerning consumers in an emerging US wine culture. Such consumers were increasingly curious and interested in good wine. Being able to consistently make such a wine enabled winegrowers to charge a high price for it. Finally, many of these new winegrowers had the requisite financial resources to sustain them in the initial years of their operation. Fixed capital costs would continue to grow in Napa, making such resources indispensable for long-term success.

During this same period, corporate interest in acquiring wineries was reinvigorated, despite earlier profitability concerns. This time, however, it was often foreign capital. Most notably, Markham Winery, St Clement, Mihaly Winery, and Whitehall Lane were all sold to various Japanese corporations in 1987, and in 1988 Kirin Brewery bought Raymond Winery. Also in 1989, Christian Brothers, one of Napa's oldest wineries and one of the big six of the post-Prohibition period, was bought by Heublein (who two years earlier had been acquired by British conglomerate Grand Metropolitan). With such purchases, corporations appeared willing to forgo earlier reservations and make the necessary investments that would not always generate immediate returns. This was a reversal of previous corporate interest in wineries. Perhaps the growth of a wine culture in the United States and abiding consumer interest in premium wines made the prospect more attractive, especially when part of a diversified portfolio. Their focus on specialist wineries rather than large ones (with the exception of Christian Brothers) suggests that, in setting their sights on niche producers, they could be more focused in their growth strategies. Such acquisitions did appear to have a positive upgrading effect on the quality of Napa wine since they brought capital that was often necessary for further improvements, and did this with a long and patient view.[23] Also, as with individual entrepreneurs, they had a long-term vision for growth and stability.

Land and its Problems

One of the results of this pattern of transactions was a renewed perception that the valley was dominated by a small number of powerful wineries. Admittedly, the larger wineries had considerable influence in many aspects of local governance, often supporting universities to develop procedures that would be beneficial to them. In terms of concentration of land ownership, it is interesting to note that in 1988 forty percent of Napa wine came from seven producers who owned 6,132 acres. However, a further 4,125 acres were in the form of leases and long-term contracts.[24] Even though the big producers undoubtedly controlled at least half of Napa's vineyards, the remaining area was dominated by independently owned small and medium-sized properties. In other words, there was much vineyard land in Napa that was not owned by wineries, and this lends credence to the continued vibrancy of niche specialists.

Earlier in this chapter I referred to the problems surrounding university recommendations of a new disease-resistant clone that proved to be anything but. While phylloxera was under control by the 1960s, its destructive potential lurked in the minds of many growers. As a consequence, viticultural specialists at universities had for much of the twentieth century (apart from the Prohibition era) been searching for rootstocks that would prove resistant to it, and offer growers reassurance. The standard vine that most growers preferred was Rupestris St George, an American vine that appeared to be impervious to the disease, and which was used extensively in the first half of the twentieth century.[25] But scientists were astute enough to recognize that the insect behind the infestation never truly went away, hence the ongoing search for an alternative and more-resilient rootstock. Several rootstocks were identified in the 1940s as having greater disease resistance, but the one that was presumed to be most effective was called AxR#1. Its proponents at the UC Davis extolled not only its disease-resistant virtues but also its vigor – something that still resonated with many growers who were searching for a vine that would provide quality *and* quantity.

AxR#1 was a hybrid developed in France in the late nineteenth century and used there extensively after earlier phylloxera outbreaks had devastated vineyards. But in the long run it failed to prove phylloxera resistant, and was extensively replaced. Similar failures occurred in other parts of Europe and South Africa. Unfortunately, and somewhat curiously, this did not deter some university scientists in the United States from continuing to investigate its viability and then promote its use, and in proffering this alternative they had a captive audience among growers. Part of the enthusiasm for replanting with this rootstock was because vineyards were aging and thus needing replanting anyway. Furthermore, since additional acreage was being planted in Napa during the 1960s and 1970s, AxR#1 seemed an obvious choice for this expansion. When growers switched to premium varietals such as Cabernet Sauvignon and Chardonnay they adopted this rootstock since it had the seal of supposed academic research approval and it also generated high yields. By 1990, approximately sixty-five percent of Napa vines had been replanted on this rootstock.[26]

Alas, unbeknownst to many (aside from the French, who presumably could have said "we told you so" had they been asked), the bug was slowly adapting to these new vines, and it eventually started destroying them. When they started to wither and die, much equivocation on the part of specialists followed, and in spite of articles in *Wines & Vines* that unambiguously pointed the accusatory finger at AxR#1,[27] it took until the latter part of the 1980s for a full recognition (and acceptance) of the problem. University specialists initially argued that other factors were at play, and when they finally acknowledged the problem they blamed the retirement of key staff and other transitional issues that blinded them from recognizing the crisis sooner (or even for advocating this rootstock in the first place). Lack of transparency followed by denial and only finally contrition did erode some of the confidence in institutional support for the industry. In some respects, this was a searing indictment of academic research at precisely the time when more in the industry had bought into the technical model of viticulture that university specialists were proposing. In the past, skepticism was unsurprising since many wineries lacked the resources to implement the suggestions proffered by specialists. That was less the case at this point, as a half-century of technical progress had produced demonstrable improvements in quality.

When wineries realized that replanting was inevitable they took solace in the fact that many vines were aging anyway (having been planted in the 1960s). Now they had a chance to plant more of the premium varietals and take advantage of clonal selections that would optimize site specificity and microclimates, and closer align with expert opinions in the 1990s.[28] The evident alacrity of this renewed enthusiasm for technical expertise is perhaps surprising given the failures in the 1970s following replanting recommendations. But notwithstanding these earlier problems, the scientific and technical approach to winemaking had become normative, and one problem would not lead to its invalidation. Too much information had been disseminated that was unambiguously beneficial, so any qualms about academic expertise would soon dissipate. Wineries were in a better financial position than in previous decades when the disease had ravaged vineyards, so they were more likely to be able to afford extensive replanting. This was some consolation, as was legislation that allowed tax reassessments for vineyards destroyed by phylloxera if they had been planted after 1978.[29] Being forced to replant, they could now choose the better-quality varietals that would afford price-premium wines. Perhaps this was the death knell for Zinfandel, Petite Sirah. and other signature grapes of earlier decades. In this respect, the phylloxera outbreak associated with AxR#1 contributed to the growing dominance of Cabernet Sauvignon as what would become Napa's signature grape.

Regulations and Identity

When Napa Valley was designated an American Viticultural Area (AVA) in 1981, it was the first in California. As noted earlier, the idea behind this classification was to provide an enhanced sense of identity for wineries in the region. Compared to similar classification in Europe, however, it was largely a descriptive tool that lacked any sort of prescriptions such as yield limits or required grape varietals. In essence, it merely constituted a classification of soil and climatic features that would be common to vineyards in a particular area.

The French Appellation d'Origine Contrôlée (AOC) system is quite different. It is a regulatory body that guarantees distinct aspects of a particular region and product. It includes soil and climatic features, but also has a set of rigorous standards for how wine is made and what grapes can be grown (and sometimes when they can be harvested). Its purpose was to replace random rules from earlier iterations and court actions with standardized procedures.[30] It is enforced by a government body (Institut national de l'origine et de la qualité), and in many respects constitutes a form of geographic protectionism. While the intent is to preserve quality (not always accomplished), it does endow wineries in a particular AOC with a certain seal of approval. This can enhance reputational credibility and is a valuable winery external resource. Similar governmental bodies provide a regulatory framework for the wine industry in Italy and Spain.

This type of top-down regulatory body, however, would clearly be anathema to American wine producers, so not surprisingly the AVA developed along less-restrictive governance lines. Even though the Napa AVA provides largely descriptive commonalities, there was the appeal of designations that would protect the overall brand, so enthusiasm for this loose classification prevailed. With reoccurring debate over the inclusiveness of the AVA, it was almost logical that, once established, the narrative might switch to further more-restrictive geographic specificity. This began a lengthy, often acrimonious and very political debate as to where boundaries should be drawn. As wineries developed in different parts of the valley, many argued that there were distinctive locational features within the overall AVA that they believed gave their wines a certain differentiating character and should accordingly be acknowledged.

In the northeastern part of Napa, for example, Howell Mountain wines appeared to express a certain character, as did the area around the Silverado Trail that subsequently became known as the Stag's Leap District. Other areas such as Oakville, Rutherford, and Spring Mountain had similar subtle distinguishing features – or so the wineries there claimed. The result of these lengthy discussions was the further breakdown into sub-appellations within the broader Napa designation. How inclusive or exclusive these regions would be was controversial, but by the early 2000s they were eventually formalized and recognized by the Bureau of Alcohol Tobacco and Firearms (BATF).[31] There are now sixteen nested appellations in Napa, varying in size from 3,300 to 16,000 acres.

Wineries felt that these were sufficiently distinctive to reflect the specific characteristics of their wine, thus allowing them a way of further differentiating their product. As more wineries adopted estate labeling for their special wines, this provided a further identity marker that permitted brand differentiation from their competitors elsewhere in the valley.

As noted earlier, not everybody was convinced that Napa should be a solely premium wine-producing area. In fact, at the opposite end of the quality spectrum, the Franzia family winery placed Napa on the label of what were bulk wines (eventually termed "super-value" wines) that were bottled in Napa but from grapes that came from elsewhere. The Franzia family had earlier owned a large bulk winery in the central valley that they sold, along with the name, in 1974. With money from these sales, younger family members established a new brand called Bronco, and then in 1994 bought Napa Creek and Rutherford Vintners. Using grapes from these wineries but also massive amounts from the central valley, the wines were made in Napa and sold for very low prices. Their wine was the quintessential commodity product, aimed at a mass market with an affordable price point. By bottling in Napa they were able, they claimed, to legitimately use the Napa designation and hence the brand recognition. Their customers might not have been fine-wine aficionados but they were familiar with Napa, so their presumption was that the wine must be good as well as cheap. We shall return to this evolving saga of the Franzia family's wines in the next chapter. Suffice it to say that the premium producers were not happy about this development as it was counter to everything they had fought against for decades, specifically removing the association between Napa and bulk wine of marginal quality.

What is a Winery?

The escapades and subterfuge of the Franzia family elicited a renewed focus on what constituted a winery. Continued growth in the number of wineries had resulted in more development and tourists, plus further interest from second-home owners seeking solace amid the bucolic charm of wine country. Napa was no longer a quiet backwater with vineyards surrounding small farm towns. It was well on its way to becoming a destination for all those who lusted after wine-related activities and an escape from the tribulations and pressure of urban life. Catering to the new tourists, many wineries subsidized their operations by offering a range of ancillary activities (concerts, weddings, cooking classes, etc.). For critics of such practices, this detracted from the agricultural essence of the valley, and was in danger of turning the area into an oenological theme park. Some winery owners on the other hand saw this as yet another activity that was part of brand and identity building, the exigency of which was becoming paramount with rising costs.

Early land use restrictions were modified in 1979 so that now the minimum lot size was forty acres (up from the earlier twenty) for a new development. This effectively rendered Napa real estate very expensive, so perhaps not surprisingly some wineries saw additional-income streams from the growing number of tourists as a desirable option. The Napa Valley Vintners Association (NVVA) supported such marketing activities, but the Napa Valley Growers Association, under the leadership of Beckstoffer, was opposed. The latter wanted a restrictive definition of a winery as a place where grapes were crushed and wine sold, and argued that a varietal specific wine (eg. Cabernet Sauvignon) under the Napa label should come from at least seventy-five percent of Napa-harvested grapes.[32] They argued that this would preserve the essential agricultural nature of the valley, but they were also protecting their own interests as growers in the face of transformations in land ownership that could disadvantage them. The NVVA did not see the developments in such a stark and damaging form, and in fact argued that such an implementation would put their members at a competitive disadvantage.[33] Each group put forward a separate proposal to the Napa County Board of Supervisors in 1987 that reflected these divergent aims.

The Board of Supervisors was the public agency responsible for land use, roads, and municipal services in the valley. The five-member board was an elected body that served a four-year term. In the early 1980s it had been largely pro-growth, but that had dissipated somewhat later in the decade with the replacement of members. Nonetheless, the election of two new members in late 1988 who were pro-growth shifted the emphasis to a continuation of wineries' existing activities. This elevated the rhetoric among those who were skeptical of the presumed benefits of future growth. What was at stake before the board was not just the definition of a winery but whether to halt the continued growth of wineries in Napa. Ironically, going along with the NVGA seventy-five percent rule would necessitate more vineyards if winery expansion was to continue. On the other hand, many felt that winery density was already growing too quickly, so further acreage even to vineyards would be problematic.

After extensive deliberation and more contentious debate, in 1988 the board introduced a seventeen-month moratorium on new winery developments, effective immediately. Then, in 1990 it passed Ordinance #947, which formalized the seventy-five percent rule, but only when applied to new wineries.[34] Furthermore, new wineries could only offer tastings by appointment, while established ones were grandfathered in and could continue to open to the public. It was a compromise ruling, and as a piece of public policy would always be subject to revision should other elected officials with different viewpoints join the board. On the whole, growers were satisfied with the passage of the seventy-five percent rule since it forced new wineries to adhere to standards that might somewhat limit growth. Existing wineries were pleased with the grandfather clause since they could continue their broad operations. Moreover, they felt that it might restrict new entrants, or at least make them less competitive if they did not have a tasting room.

Overall, the rulings revealed the ongoing tensions between pro-growth groups and those whose vision of Napa hearkened back to its agricultural past. It also pointed to the fragility of rulings that could easily change depending on the balance of power in such public bodies and the ability of the various factions (including developers and existing landowners) to exert influence over the election of such members. Arriving at an accepted regulatory framework for winery operations and the potential for expansion would figure prominently in future public-policy hearings in the decades to come. In most cases, it was a struggle between aspiring newcomers, desirous of entering the industry, and established groups who felt that the valley was already too crowded. Density, as we have seen, was incredibly important for early industry growth. The question now was whether enough was enough.

The Napa Style

As Cabernet Sauvignon and Chardonnay emerged as the two predominant grape varietals, so too did discussions about the stylist characteristics of wine from the valley. Vineyard acreage continued to expand during the 1980s, with much of the new acreage devoted to these two varietals. While not necessarily cash cows, they were vital components of a revenue stream predicated on leveraging the premium pricing they offered. Cabernet Sauvignon was increasingly recognized as the most profitable varietal, and by 1989 Bordeaux varietals (including Cabernet Franc, Merlot, and Petit Verdot) were the majority acreage in the valley.[35] Demand for such reds, in which Cabernet Sauvignon was the predominant grape, was steadily increasing. Since quality had significantly improved over the past three decades, and alongside a rise in these grape prices, it appears that consumers were willing to pay more for such wines.

Evoking, even implicitly, the link with Bordeaux, and despite many wineries purportedly benchmarking their wines against the best from Bordeaux, stylistically Napa reds were quite different from their Old World cousins. In the past, the similarities might have been nuanced but certainly evident. Now, many wines from Napa possessed a unique style that clearly distinguished them from their Old World counterparts.

As noted earlier, the growing season in Napa is longer, hotter, and drier than in Bordeaux. Under such conditions, it is easy to ripen Cabernet Sauvignon. In fact, it became increasingly recognized that not only would ripening be fairly straightforward, but the grapes could be left on the vines longer than in Bordeaux to provide a more enhanced flavor profile. The rationale behind this came from studies suggesting that physiological ripeness would improve the eventual quality of the wine. Thus, more wineries concentrated on achieving optimal phenolic ripeness, which is the best indicator of the inherent properties of a grape's development. But, since phenolic ripeness often lagged behind sugar developments, the resulting wines tended to have a higher percentage of alcohol, and were described as rich, fruity, or even jammy in their profile.[36]

As more wineries adopted this style of wine, other processes were introduced to moderate tannin levels to make the wines more readily accessible (without the requisite aging). These included extended maceration, whereby fermented must is held in sealed tanks for a longer period than normal, forcing bitter tannins to fall to the bottom. The resulting wine does have a nice balance, despite its power, and is a richer, deeper red, and noticeably flavored (sometimes even appearing to be slightly sweet). By the late 1980s, this style of winemaking was increasingly the norm in Napa.[37]

It could be argued that such a profile appeared to be popular with consumers, especially when some of the small production, limited-availability, and high-cost wines developed that style and immediately sold out. Others have argued that the success of such wines came because they stood out in blind tastings when critics lauded them with high scores. Consumers saw this formal approval as an affirmation of the quintessential quality of the wine. Some critics, such as English wine-writer Jancis Robinson, who do not favor such a style have claimed that these red wines "muscle out finesse."[38] In fact, Robinson has consistently argued that such wines simply overwhelm subtler others in tastings. For their part, producers have often countered that such wines are merely an expression of the vineyard and the conditions under which the grapes were grown that clearly favored such a wine type and profile. In an effort to mute some of the criticism, some went so far as claiming that such wines were low in tannins and therefore more immediately accessible, unlike Bordeaux wines that needed cellaring. Since most buyers consume wine shortly after purchase this has proved a desirable hedonic feature for them.

In the past it was assumed that Napa wines would continue to improve with cellaring. This new style of what some referred to as overripe grapes obviated the need for such aging, making the wines immediately accessible to the new wine connoisseurs. Would such wines change and even improve with age? That was the question that even some winemakers were unable to answer. With the continued sanctification of such a profile by critics, many wineries that might have eschewed such conformity in the past saw it as disadvantageous to be too much of a stylistic outlier. If consumers continued to purchase such wines – even if they did not always trust their own instincts and instead preferred the credibility conferred by critics – on their part wineries continue to aver that their wine is a natural expression of place and see no reason to change their practices. As more wineries coalesced around producing this style of wine, it would become one of the distinctive features of Napa Cabernet Sauvignons with a profile resonates with some but garners concern among others.

Notwithstanding such stylistic convergence, it is important to note that some wineries chose a different path. Iconoclasts such as Cathy Corison have eschewed such approaches. Corison was one of the early winemaking pioneers in the valley, working at numerous vineyards before establishing her own label in the late 1980s. A professionally trained winemaker, she was familiar with different approaches to viticulture and viniculture, but preferred to make a more restrained wine. She abstained from the above harvesting practices, picking her grapes relatively early to optimize what she felt was the right balance between sugar and acid levels. As a consequence, her wines were lower in alcohol (13.5 to 14 percent), well balanced, and with greater aging potential. Randy Dunn had a comparable approach. Both produced wines of exceptional quality at price points that firmly placed them in the ultra-premium category, but they are noticeably different from what many think of as Napa Cabernets. They continue to receive wide acclaim, even from critics, but this raises the question of the true expression of place in a wine, and how much differentiation there might be between adjacent areas. This returns us to discussions not just of terroir but also the technical mediation that goes into winemaking, and what this means for overall identity. Napa's evolving reputation was predicated on certain grapes and increasingly a style or character of the wine made from them. Precisely how much variation can one sustain without undermining the fundamental coherence of this organizational form or cluster of wineries? This is the question of the ensuing decades that we turn to in the next chapter.

Conclusion

By the late 1980s there were more than two hundred wineries in Napa. Many were small with a few thousand cases per annum, while others were large with over one hundred thousand cases. This multi-decade work in progress that culminated in the Paris Tasting led to the acknowledgment of Napa as capable of producing excellent wines. Individuals intent on starting a winery continued to be attracted to the region, increasingly bringing the financial resources and capabilities necessary for a capital-intensive operation where revenue streams are inevitably postponed for several years. There was also a growing number of professionally trained winemakers who provided the formal knowledge and skillsets to consistently produce a quality wine. Many of these winemakers trained at UC Davis, where research continued to offer a range of solutions to routine problems. Some have argued that such research was directed more at large wineries where technical proficiency was designed to complement such scale. Winemaking had become much more of a science than an art, and all too often winemakers became risk averse. Be that as it may, we have also seen the powerful instructional role of informal education and the sharing of tacit knowledge through groups such as the Wine Technical Group and their monthly meetings. This emerging community of similarly-skilled individuals permitted collective-organizational learning and problem solving in a cost-efficient and expeditious way. Admittedly, formal practices were constantly being developed in areas such as planting density, clonal selection, trellising systems, and ultimately how best to balance quality and yield. But it was the synthesis of technical education and informal knowledge transfer that enabled wineries to capitalize on the area's unique characteristics.

For their part, winery owners found a collective voice in the NVVA. Through this organization as well as the myriad of informal links they had established with each, they were able to more clearly articulate Napa's identity. Most had realized the quality of Cabernet Sauvignon as a premium varietal that not only grew well in various locations, but could in their winemakers' hands produce a stylistically distinctive wine that many critics favored. In developing the configurations of a market for fine wine in which the singularity of their product engendered authenticity, they were able to differentiate theirs from Old World wines. Although, locationally and culturally, they were part of a broad wine market, they had engineered a distinctiveness by acquiring and developing somewhat unique operational techniques. This form of collective organizational learning, in which they could develop and protect their emerging brand identity, would be crucial for their continued reputation building. This didactic approach to routine activities embellished the status of their product and enabled them to differentiate it from other regions. In fact, the putative difference in the style of Napa wine rendered it unique in the eyes of consumers, who conflated the rising prices with presumed quality attributes. It has often been said that anybody can make wine but not anybody can make fine wine. To be accomplished at the latter, the maker needs skills, resources, and the right location. More and more of the new winery owners personified these attributes as they recognized the imperative of an unwavering commitment to quality and the uniqueness of Napa that made such aspirations possible.

American wine consumers were also changing their habits. Wine had become demystified for many, and those with greater familiarity – and increased discretionary income – were willing to pay a price premium for Napa wines. As a status good, wine continued to be a marker of success. The more expensive the wine, the better the presumed quality. Limited availability (the scarcity factor) further added to the prestige, and since many of the new wineries had very limited production they could use pricing as a signifier of quality. The essence of this is succinctly expressed by a winery president who stated, "Napa is able to charge a lot because there are a finite number of places in the world to grow great wines but an increase in demand for them. People are prepared to pay a premium for the product because of status, feel good factors and quality. Production changes over the years, a better understanding of viticulture and the ability to consistently make a great wine are features that have driven Napa's success over the years. But it also important to keep driving up the prices and limit selection to add to the mystique."[39] The Paris Tasting demonstrated that Napa possessed the attributes for excellence. Those who entered the industry in the 1970s and 1980s had the capabilities to garner resources in ways that systematically and more consistently made such excellence possible. The question was whether or not such trends would continue if more wineries were founded. Would increased density lead to less cooperation and more competition, leaving the role of niche producers more vulnerable? Might institutional forces and public policy constrain growth in ways that limited competitive advantage? Or would consumer demand continue to grow as the region's reputation became more established, and new customers with greater discretionary income became enthusiastic wine buyers? We examine these issues in the next chapter.

Notes

[1] James Lapsley, *Bottled Poetry* (Berkeley: University of California Press, 1996), 200.
[2] Matt Kramer, *New California Wine* (Philadelphia: Running Press Books, 2004), 72. See also Jacques Delacroix and Michael Solt, "Niche Formation and Entrepreneurship in the California Wine Industry 1941–1984," in *Ecological Models of Organizations*, edited by Glenn Carroll (Cambridge, MA: Ballinger, 1989), 53–70.
[3] For a comprehensive discussion of this concept and status in general see Cecilia Ridgeway, *Status: Why is it Everywhere? Why Does it Matter?* (New York: Russell Sage Foundation, 2019).
[4] Kramer, *New California Wine*, 29.
[5] Charles S. Sullivan, *Napa Wine* (San Francisco: The Wine Appreciation Guild, 2008), 41.
[6] The idea that social practices are responsible for regional dynamism has been developed by economic geographer Michael Storper in his analysis of industrial district growth. See Michael Storper, *The Regional World: Territorial Development in a Global Economy* (New York: Guilford Press, 1997). See also Greig T. Guthey, "Agro-industrial Conventions: Some Evidence from Northern California's Wine Industry," *The Geographic Journal* 174, no. 2 (2008): 139.
[7] See Anand Swaminathan, "Resource Partitioning and the Evolution of Specialist Organizations: The Role of Location and Identity in the U.S. Wine Industry," *Academy of Management Journal* 44, no. 6 (2001): 1169–85.
[8] See, for example, Anand Swaminathan, "The Proliferation of Specialist Organizations in the American Wine Industry, 1941–1990," *Administrative Science Quarterly* 40 (1995): 653–80; Glenn Carroll, "Concentration and Specialization: Dynamics of Niche Width in Populations of Organizations," *American Journal of Sociology* 90 (1985): 1262–83.
[9] For example see Stephen Brook, *The Finest Wines of California* (Berkeley: University of California Press, 2011); Steve Heimoff, *New Classic Winemakers of California* (Berkeley: University of California Press, 2008).
[10] Doug Shafer, *A Vineyard in Napa* (Berkeley: University of California Press, 2012), 116.

[11] Personal interview, 2005.
[12] Sullivan, *Napa Wine*, 331.
[13] James Laube, *California's Great Cabernets* (San Francisco: Wine Spectator Press, 1989), 154.
[14] In 1933 Jack Riorda and some partners (including Cesar Mondavi, Robert Mondavi's father) built a winery named Sunny St Helena. It was sold in 1946, whereupon it became a cooperative winery, and in 1972 was sold again to the Christian Brothers who used it as a storage facility. Bill Harlan and his partners purchased it in 1986, and in 1991 the name was changed to Merryvale. In 1996 one of the partners, Jack Schlatter, became the sole owner.
[15] Personal interviews with Bill Harlan, 2019 and 2020.
[16] Kim Marcus, "Bill Harlan's Amazing Saga," *Wine Spectator* (November 15, 2015), 61.
[17] James Laube, "Master of Style," *Wine Spectator* (November 15, 2015), 64.
[18] Laube, "Master of Style," 61.
[19] It is worth noting that one of John Daniel's daughters, Robin Lail, who inherited Inglenook worked for Bill Harlan in 1982 at his real-estate firm, and during her time there and afterwards provided him with ideas and even advice that came from a uniquely insider's view of Napa. See Mary Anne Worobiec, "Creating her Own Legacy," *Wine Spectator* (May 31, 2020), 40–50 for further details of their relationship, and that with Robert Mondavi.
[20] Personal interview, 2006.
[21] Ibid.
[22] J. W. Rinzler, *Daring to Stand Alone: An Entrepreneur's Journey* (Petaluma: Cameron + Company, 2018); personal interviews, 2017 and 2018.
[23] Sullivan, *Napa Wine*, 363.
[24] Ibid., 362.
[25] Ibid., 368.
[26] Ibid., 373.
[27] *Wines & Vines* (February 1979), 52–4.
[28] Sullivan, *Napa Wine*, 373.
[29] Ibid., 373.
[30] See Rod Phillips, *French Wine: A History* (Berkeley, University of California Press, 2016), 238–43. The introduction in 1935 was a modification of earlier rulings which had been controversial for those wineries who were omitted from the classification. A similar set of concerns followed this ruling, but it is notable for its expansive definitions that went beyond terroir narratives.
[31] One somewhat acrimonious debate surrounded who could be included in the new Stag's Leap District. A restrictive definition would allow wineries to the east of the trail to be part of the appellation, but new wineries such as Silverado and Mondavi's family lands to the west wanted to be included as they argued the soil types were similar in both areas. The eventual ruling by BATF in 1989 accepted the expansive definition that included both land to the east and west of the trail. See Sullivan, *Napa Wine*, 355–9.
[32] Sullivan, *Napa Wine*, 281
[33] Ibid., 348.

[34] Ibid., 281.
[35] Ibid., 386.
[36] Ibid., 389.
[37] *Wine Business Monthly* (September 6, 2003).
[38] Quoted in Sullivan, *Napa Wine*, 390.
[39] Personal interview, 2005.

CHAPTER SIX

AFFIRMING THE NEW ORTHODOXY

The 1990s were the period of the dotcom boom that brought more new owners to Napa, as well as new customers for Napa wines. The former were often resource-rich individuals looking for a lifestyle change as well as those who firmly believed that this was the time to realize their oenological dream by acquiring land. They had made fortunes elsewhere, appreciated fine wine, and seized the opportunity to buy into the emerging Napa dream. Their properties have sometimes been disparagingly referred to as "trophy wineries" for people who like the hype and glamor of such ownership. In some cases that is undoubtedly accurate, but for others it is axiomatic that their desire to craft a truly excellent wine was the motivating goal. They brought a vision, commitment, and, perhaps most saliently, abundant resources to make that goal possible. There is no reason why their acquisition couldn't simultaneously be a status good and an operational embodiment of excellence.

Those not wishing or able to buy a winery became avid consumers. While overall wine consumption declined in the United States during the early 1990s, demand for Napa wine continued to grow. Specifically, after 1993, purchases of wines priced fifteen dollars and above increased, which suited Napa given its premium pricing strategy and growing reputation. People were drinking more and better Napa wines, despite the downturn in other sectors of the market. Evidence of Napa's credibility can be seen in the ratings of Cabernet Sauvignon in *Wine Spectator*. Between 1988 and 1993, the average score was 89.6; between 1994 and 2001 it was 94.[1] This affirmation by respected critics, and its impact on sales, is testament to the continued improvements in quality as well as the growing role that critics had in legitimizing that quality. For the uncertain wine consumer, the route to connoisseurship was eased through such ratings, especially for wines that were more costly.

As the number of wineries continued to grow during this period, it is important to note that the majority were still small and family-run, some with a long history in the valley, and many being the pioneers of the 1970s. Annual production was typically around fifteen hundred to ten thousand cases. Some used a consultant winemaker, but most employed winemakers who had trained at UC Davis (or in some cases Europe). Among this group of small wineries, however, was a new category of recent industry entrants that was attracting attention. These comprised newcomers who brought with them extensive financial resources, enabling them to develop operations but with less constraints on immediate profitability. Notably, they often reshaped many of the operational norms, sometimes eschewed the consensus building that mediated owner-government relations (especially over land use), and embraced a business model of small production, high price, and limited availability. They also helped cement the brand reputation of Napa, in part by drawing attention to their own often-public endeavors.

This category of new owners was somewhat different from those of previous decades, inasmuch as they not only brought considerable financial wealth but were clearly prepared to forgo profitability for much longer than accepted wine-business practices. Wine production has rarely been an industry in which revenue streams can quickly lead to profitability. In fact, the opposite is generally the norm. While white wines can be brought to market more quickly, reds require aging that can add several years to production. Stated simply, whites can conceivably be seen as cashflow while reds are inventory – with all of the costs associated with that. Since Napa increasingly enshrined Cabernet Sauvignon as the preferred varietal, the time to market (and positive revenue streams) became extenuated. Attaining profitability under such conditions would always be hedged with uncertainty and delay.

Scott Morton and Podolny, in their assessment of ownership trends in the California wine industry, note how many wealthy individuals entered the industry with a focus on a broad definition of utility (lifestyle, ownership enjoyment, status, etc.) rather than profitability per se.[2] This wasn't necessarily a new phenomenon, but it took on a more pronounced form in the 1990s and contributed significantly to Napa's continued growth. The organizational framework that ensued both re-affirmed technical proficiency and modified extant cultural practices of collective learning. New owners were patently a somewhat different breed, and were willing to do things differently. In time, this resulted in modifications to the way networks worked in Napa – an abridged version of the cooperative structure introduced more competition among firms.

For many new firms, their commitment to maximizing operational efficiency and producing the highest quality wine resulted in enhanced productive services, both within the firm as well as industrywide. In other words, it was not just a question of resource possession but the services that such resources render dynamic. Being able to identify and operationalize key dimensions of the winemaking process (from vineyard all the way to the cellar), and to do so by marshalling the requisite components in the pursuit of excellence, enabled them to attain high levels of quality. This resonates with Edith Penrose's earlier articulation of firm dynamic capability in which the manner that resources are utilized and productive opportunities are developed is unique to firms, but also contingent upon technical parameters and network interactions between similarly-endowed firms.[3] But in the case of Napa wineries, utility maximizers' unwavering commitment to excellence and their ability to deploy resources accordingly raised the bar for incumbents, many of whom were by necessity strictly profit maximizers. Now, to survive competitively necessitated a similar approach for the latter, which resulted in enhanced resource-creation activities for the industry as a whole. Such activities raised the bar for earlier incumbents, although it was precisely this group that had set in motion the systematic upgrading of quality and credibility that attracted the current crop of newcomers to the industry. Despite these changes, and while no longer an entirely collaborative project, the collective organizational learning that continued to develop was contingent on capability specialization, technological experimentation, and a de facto normative framework for improving quality.

The more successful incumbent firms became, the more newcomers were attracted to Napa. The increasing density of firms and the embedded relationships between them resulted in a further accumulation of knowledge that allowed firms to differentiate and focus on their core capabilities without compromising the broader network efficiencies.[4] Furthermore, the dense sets of relationships and reciprocal obligations were mutually beneficial for firms, and their inclusiveness (plus informality) did not privilege knowledge transfer that often occurs in more hierarchically structured markets. At least, this was the case initially. It gradually changed as new firms became established, however. Such knowledge was the dynamic resource that facilitated the creation of new productive resources, which ultimately drove economic growth in the sector. The vast repertoire of site-specific practices that constituted tacit knowledge was already being shared informally, as well as under the guidance of institutional bodies such as the NVVA and the NVGGA, which encouraged and often systematized such endeavors. It now took on more urgency as additional newcomers sought participation and inclusion (plus a shared vision), but also brought and bought skills, expertise, and techniques designed to further enhance the credibility of the industry.

Cabernet and Quality

It would be disingenuous to claim that Napa's success during these decades came entirely from Cabernet Sauvignon. Chardonnay had become the dominant white varietal planted, and Sauvignon Blanc would eventually be popular. Some growers still stuck with Zinfandel while others were finding success with Pinot Noir. Petite Sirah still had a limited following, but the old-time grapes such as Alicante Bouschet were becoming harder to find. In this new era, they were simply not profitable to grow. But if one looks back to 1961 when Cabernet Sauvignon accounted for a mere six percent of the total acreage, it was dramatically different from now where it is just over fifty percent. Bordeaux varietals had come to dominate the grapes grown in Napa, and increasingly they were what people associated with Napa wine. To get a better sense of just how important these varietals had become, it is worth looking at acreage trends.

Comparing 1994 and 2006, one finds Cabernet Sauvignon accounting for 10,209 acres in the former and 18,883 in the latter. Merlot went from 3,804 acres to 7,042 in that same period; Cabernet Franc from 751 to 1133; and Petit Verdot from 87 to 543.[5] The latter was a popular blending grape in Bordeaux, where it added color and density to the wine. However, it was never a standalone varietal (as were the first two), so its small acreage understates its significance during these time periods. Between 1993 and 2006, Chardonnay acreage increased from 4,425 to 6,749, while Chenin Blanc dropped dramatically from 2,167 to 38, as did Riesling from 1,330 to 138. Sauvignon Blanc remained approximately the same during this period, going from 2,040 to 2,025 acres.[6] Clearly there was less interest in white varietals than Bordeaux reds.

As noted in chapter six, much of the Cabernet Sauvignon increase came as a result of the replanting following phylloxera and its potential future threat. But these new vineyards also capitalized on innovations in viticulture as well as rootstocks, and the result was a significant modification of how the vineyard looked and was managed. The aim for many winegrowers was to suppress the vigor of the vine and get lower yields.[7] This might seem counterintuitive for an agricultural crop except that growers found that lower yields increased the flavor intensity and quality of the grapes. Furthermore, those hillside vineyards that had become increasingly popular found that their thin soils appropriately stressed their vines and gave them the requisite quality.

Alongside the quality improvements came a continuation of the stylistic changes that were discussed earlier. Riper fruit as a result of phenolic changes gave rise to richer, fuller-bodied wines. This is why so many Napa wines are referred to as "fruit forward," because there is a considerable "jamminess" and richness in the wine, plus an increase in alcohol levels because of the enhanced sugar levels.[8] Sullivan notes that this trend accompanied changes in the cellar when extended maceration was used to lengthen the fermentation process, which results in tannins sinking to the bottom and subsequently being drawn off.[9] Absent excessive tannins meant wines could be consumed younger, as one did not have to cellar them for years to make them approachable. Tannins generally provide age worthiness, and until recently were the norm in France. But American consumers appeared to want a wine that could be drunk upon release without compromising its profile and taste. These viticultural and winemaking techniques delivered such a wine, albeit one that was increasingly notable for its richness. In fairness, many winegrowers claimed that such a style was the natural expression of their vineyard characteristics – winemakers were merely agents in that process, albeit in ways that could modify the outcomes to develop a certain profile. Is this what consumers wanted? A Faustian bargain to gain immediacy from wine at the expense of a putatively more balanced drink with lower alcohol levels – the proverbial food-friendly wine? Given the increased consumption, the answer is a tentative "yes."

At this juncture, it is important to recognize the decisive role played by critics and their endorsement of this stylistic change. The most influential (Robert Parker's *Wine Advocate* and *Wine Spectator*) lauded wines with such a profile. Their importance as de facto gatekeepers increased during the 1990s as new consumers sought validation for their purchases, allaying their insecurity or lack of knowledge. With prices rising, consumers wanted reassurance that an expensive bottle was justified by its presumed quality. The high scores that the critics gave (as well as extensive written comments that were sometimes read but often ignored) affirmed the price/quality ratio.

Not all wine critics appreciated the new Napa Cabernet style, and in fact some, such as English writer Jancis Robinson, derided it. But those that did extol its virtues provided a solid dose of legitimacy for wineries that could now more consistently produce this type of wine. I will return later to the important role of critics and the extent to which wineries possibly make wine that unambiguously caters to such tastes. Suffice to say that consumers appreciated this style in sufficient numbers to drive the prices up, making Cabernet the obvious varietal to grow since it commanded a price premium. The mean price of a 1994 Napa Cabernet vintage was $27; by 1999 it was $44.50. With some minor fluctuations, it has increased since then at a rate that far exceeds inflation. Admittedly, grape prices increased during this same period, but across all varietals at a similar rate. For growers, however, especially the NVGGA, it became clear that this had become the signature grape, and when grown according to the new viticultural norms it could command a high price. In fact, they arrived at a pricing formula whereby grape prices were charged as a ratio (one hundred to one) of the final retail bottle price.[10] Negotiations between growers and wineries eventually formalized this arrangement, so at least growers were able to benefit financially from the growing popularity of Cabernet Sauvignon.

As quality continued to improve and a distinctive style of Cabernet that could be sold at a price premium developed, it increasingly became the region's signature grape. Perhaps not surprisingly, some wineries refocused their production on this grape. One owner told me how they had ripped up their Sangiovese vines and replanted with Cabernet Sauvignon. The former had produced some excellent wines, but Cabernet could be sold at three times the price, making the Italian varietal economically far less viable.

The Cult Wine Phenomenon

Cabernet's success was also writ large by many of the new smaller wineries that were opening since it was the focus for their production. With a dedicated commitment to small-batch production (often less than fifteen hundred cases annually) and an assiduous commitment to quality, they helped solidify and refine Napa's reputation. Their virtually exclusive focus on red wine (specifically Bordeaux varietals) both exemplified the valley's viticultural excellence for these grapes and reinforced the core competence of the producers.

One of the best known of this type is Screaming Eagle. In 1986, former real-estate agent Jean Phillips bought a fifty-seven acre vineyard near Oakville that was planted to several varietals. She sold grapes from her vineyard to local wineries but kept a one-acre plot of Cabernet Sauvignon to experiment with making her own wine. In her search for professional staff, she discussed options with various winery owners in the area before hiring Richard Peterson as a consultant, and then his daughter, Heidi Peterson Barrett, as a winemaker. She released her first vintage (1992) from this small plot in 1995, and received rave reviews from wine critics who awarded it exceptionally high scores (99 from Robert Parker). The entire vineyard was replanted in 1995 with Cabernet Sauvignon, Merlot, and Cabernet Franc. In 2006 she sold the winery to billionaire Stan Kroenke and Charles Banks for an undisclosed fee, and in 2009 Kroenke became the sole owner.[11] The current annual production is four hundred to seven hundred and fifty cases. The wine is sold on allocation and retails at approximately three thousand dollars a bottle.

Similar to the above in strategy and operation are the Bryant Family Vineyard (est. 1987, first released vintage 1992) on Pritchard Hill; Colgin Cellars (est. 1992, first vintage released in 1995) also on Pritchard Hill; Araujo Estate Wines (est. 1990 when Bart and Daphne Araujo bought a 160-acre property near Calistoga that included the famous nineteenth-century Eisele Vineyard); Schrader Cellars (est. 1998) that doesn't own vineyards and contracts with grape growers; and Dalle Valle Vineyards (est. 1982, first vintage released 1988) in Oakville, where twenty-one acres are planted to Cabernet Sauvignon, Cabernet Franc, and Petit Verdot. Most produce around two thousand cases annually, generally sold on allocation and at price points above three hundred dollars a bottle. All have been sanctified with high scores by the critics.

Such wines have been labeled "cult wines" because they hew to a very specific production strategy – limited production and availability, high prices, an almost compulsive attention to detail to produce an excellent wine. The best winemakers and vineyard managers are hired, and famous consultants (many from France such as Michel Rolland) are often used to provide further guidance. By adopting a very high premium-pricing strategy, they increased the desirability of their wines. And because of limited availability, their wines acquired an aura of exclusivity that further enhanced their value.

It is wineries such as these, in addition to several mentioned in the previous chapter such as Harlan, that have come to be seen as the new "luxury" face of Napa wine. In fact, as many strive to replicate the reputation of Old World "first growths," they are purposefully consolidating the credibility of Napa wine as well as leveraging that status to differentiate their brand from others of similar ilk. Their explicit aim has been to produce an estate-demarcated wine in small quantities and at a high price. Admittedly, their size often precludes large production, but the pricing strategy nonetheless sends a message to consumers about their quality intent. Many benchmark Bordeaux first growths, not just in the quality of the wine, but also as a basis for these pricing strategies. Cognizant of the link between price and perceived quality, they are attentive to what they feel the market can bear. According to one winery owner:

> You charge what you think the market considers a fair price for a really fine wine. Customers perceive a relationship between quality and price. The higher the price, the higher the status of the wine, especially if it receives a high score. Our goal was to make a really fine single-estate wine and deliver it with infallible consistency. Bordeaux was our model, but since we don't have a track record that they have we had to find other ways of signaling our value. That's where Jim Laube [Napa critic for *Wine Spectator*] and Parker came in, especially at the beginning when we were trying to build up a select mailing list. Now we don't worry so much about scores.[12]

Another owner was even more explicit about his intent, taking deliberate aim at a small segment of a high net-worth individual to whom he could sell his wine:

> I wanted to make a wine that could be a collectable asset that would appeal to the richest one percent of the population. My initial investments were very high, but I covered the costs from my other businesses. This was about creating a new product in this vibrant market; making something that would stand out from the others. One way of doing this is to set a high price at the beginning. I chose price points for my Cabernet that were about forty percent higher than that of established wineries in the areas and went from there. If you start high it sends a message that you are raising the bar. Given the publicity that surrounded each venture like ours, people seemed willing to accept that this might be a fantastic product. In a way, I suppose I let rich people determine my prices – the higher the prices, the more likely only a few could afford it, and this confers a status on those that can. They like to feel as if they are in a select group, and this is the importance of a select mailing list.[13]

An owner of a twenty-five thousand-case winery said his aim was to consistently produce a quality wine that was worthy of comparison to a French first growth. To be able to do that repeatedly is crucial because ultimately it affects the image of his wines. Since the winery is marketing according to luxury-product rules where image is so important, that had to be based on managing a consistent high quality. This enabled him to charge a high price for the wine. He said: "our pricing reflects this ... if you don't charge a lot you don't get noticed."[14]

The issue of pricing is important since winegrowers have more latitude in setting prices than occurs in other markets, where intermediaries have greater influence. In this respect, they are price makers rather than price takers – in other words, they possess greater capability to set their prices according to their own market evaluation rather than be subject to external constraints. Such a pattern approximates that of Bordeaux first growths that are able to exercise similar pricing power while other premium producers in that region remain at the mercy of coordinated market mechanisms in the Place de Bordeaux (the formal market intermediary for wine sales from that region).[15]

Most of the winery owners I have interviewed over the years claim that pricing is a crucial part of the overall packaging of their product. It is indispensable to claiming a status ranking and resonates with consumers who are looking for some verification of quality. Aside from critics, pricing is a quintessential signifier. As one owner told me, "nobody looks for a bargain in a luxury wine – at least they won't admit it. In fact, quite often the more they pay, the more secure they feel in what they have purchased."[16] The marginal utility remains high and becomes higher when the price increases. Another owner commented that "packaging, price, product and story" are all crucial elements of success, "but price gets your name out there immediately."[17] Opus One demonstrated the efficacy of such a practice when its initial offering was set above the average prices. Doing this announced the intent in a very dramatic way, and garnered publicity that could be sustained, but only if the quality measured up to the hype. Winegrowers can often justify high prices by claiming high fixed costs (land, winery operations) as well as high variable costs (winemaker, consultants, etc.). But most acknowledge that to deviate from a high price strategy would be counterintuitive given the valley's growing reputation as a producer of ultra-premium wines. Very few wineries appear to compete on the basis of cost, instead seeking subtle ways of brand differentiation that rely on stories, traditions, practices, and site specificity.

Limited quantity or scarcity of supply is another key ingredient. Since most of the wineries have limited acreage under vine, output will inevitably be limited, especially if they pay scrupulous attention to selecting only the best fruit. Some might like to expand production, but the absence of land precludes that option. Most wineries invoke the notion of artisanal quality, a craftlike approach to production where highly skilled individuals approach operations with meticulous attention to detail. Labor-intensive production (handpicking grapes and hand sorting were deemed essential) was the norm. Give that most winemaking is embedded in a techno-scientific paradigm, reconciling that with traditional techniques bridged old and new. Furthermore, it enabled wineries to invoke tradition when in fact they had none. It provided a suitable umbrella of heritage that enveloped the otherwise contradictory practices of traditionalism and modern techniques. It also gave credence to their growing assertion of site specificity, even if it was not evoked with the same intensity as the French debates on terroir. Here are the words of one owner of a limited-production estate:

> Our goal was to make a series of single-estate, limited-production, world-class wines. We had great hillside locations and fantastic soil. We aimed to produce several hundred cases from three estates. We built a small mailing list with information from restaurant owners, wine merchants, and people we knew who were passionate about fine wines. Our first release was limited to three bottles per customer. Most wanted more, perhaps a case or two, but we had neither the capacity nor the inclination to meet this demand. We created an immediate perception of scarcity which has helped us develop our reputation. The harder it is to get, the more people seem to want it. The biggest challenge has been saying "no" to very wealthy individuals who are used to getting what they want.[18]

Another owner said:

> What we are doing is an artisanal business that is selling an experience. Once you've drunk it, it's gone, so much of the value comes in the anticipation. In many respects, we are the least businesslike in what we do because we're trying to limit sales. So what wine we sell is basically the experience of being part of a bigger project that has to be expensive. The average person doesn't get it, and that's fine. They're not our customers.[19]

Winery/Consumer Interface

If you break winery operations into three components – growing the grapes, making the wine, and selling the wine – which is the most difficult? This is a question I frequently posed to winery owners. Since most were making very high-end wine and, apart from a very few with extensive waiting lists for their allocation, the answer was typically the same – selling the wine. One owner of a winery that figures prominently in auction indexes said: "I spend much of my time on the road, doing wine dinners, talking with distributors, doing small events. It's constant. But if I don't do it the wine might not sell."[20] Many of the small wineries who sell direct to consumer typically have more extensive information about who consumes their wine than most producers, who rely on distributors or a wine club with many thousands of members. Knowing their consumers is important since it enables them to build relationships and often personalize interactions.

For the winegrowers that maintain allocation systems, this provides a further exclusivity dimension. Customers on such lists see themselves as being members of a privileged club for which the price of entry is very high, and thus exclusive. When asked "how many people don't take up the allocation offer?" the typical response was "very few." And the fact that there are long waiting lists before one secures an allocation further demonstrates their resilience. For wineries, this ensures that the vast majority of their wine is effectively presold, with a small amount perhaps going to restaurants and a few key distributors as part of brand building. The longer one is on the list to receive wine, the more it enables you to drink down older vintages. New vintages to be laid down replace older ones that are consumed. This renders ephemerality less consequential, since an individual's cellar remains a constant repository in terms of quantity, even if there is turnover in actual bottles.

The relationship between winery and consumer is paramount since it removes much of the opacity and even uncertainty that can surround wine purchases. One owner of an early cult pioneer commented

> Our aim from the beginning was to make high-quality wines that would appeal to wine enthusiasts who appreciated a finely crafted product. But we also spent a lot of time focusing upon customer relations and customer service. It was important for us to know who was buying our wine, why they liked it, and what their overall experience was. It's easy to just sell wine and ignore consumers as long as they continue to buy. But we really believed it was important to constantly deliver a level of service and response that would not only nurture buying but also make them feel that they were part of the whole operation. I believe this is part of the reason we have been so successful for so long with so many repeat customers.

When asked about who specifically buys their wine, whether from mailing-list details or general information gleaned from retailers and distributors, owners were fairly consistent in their responses. Clearly, it is a small subset of the population – the proverbial "one percent," as several commented. Others were more specific and said mainly doctors and lawyers, and in recent years sports stars and those in the financial industry. The dotcom boom that heralded the growth of boutique-style wineries during this period also swelled the ranks of consumers with notably discretionary income. As to the question of why people buy their wines, some owners felt it was for bragging rights ("I have x number of ninety-five-plus wines in my cellar"). Proudly displaying one's cellar and its contents can be a form of one-upmanship, or it might just be an indicator of pride at one's possessions. Certainly, the increased publicity for the aesthetic dimensions of wine cellars (such as those featured in *Wine Spectator*'s "celebrity" cellars) reinforces the glamor and sophistication of wine ownership. Having the appropriately sanctioned wines in such cellars becomes de rigueur for some wealthy wine enthusiasts. One owner went so far as to distinguish two sets of his customers in a succinct and highly descriptive way: "wine geeks and stamp collectors." The former is typically a knowledgeable oenophile, while the latter characterizes the individual who collects only wines with the highest scores that they can display to their friends. The latter behavior is certainly consistent as a status marker for a Veblen good, and presumably conjoins the various sensory dimensions of the wine with the material pleasures that come from possession. It could also be said that this was a way for the rich to demonstrate just how rich they had become. Wine, like art, had gained currency, and it was now almost an emblematic capital asset.

As repeatedly noted, critics play an essential role in these transactions as an intermediary between the wealthy-but-uncertain buyer and winegrower. Typically, they confer the scores that the latter use to justify their purchases. One winery owner said that some buyers are insecure and look for heuristic markers to judge quality, and this is precisely what critics make possible. Another owner of a small winery said critics' high scores were essential for his sales, and those of other smaller wineries:

> Big producers have less need for critics since they have the muscle to get their products out there, but since America is not a wine-drinking culture, people need educating and help in this area. Also, we are children of soundbites, we have short attention spans, and an unwillingness to investigate in detail about a product. The average consumer doesn't have much time so needs information and critics provide this. This is where Parker and *Wine Spectator* matter, since they inform people but their scores can make or break a winery.[21]

A high score or being named "wine of year" by *Wine Spectator*[22] can generate a huge increase in demand, even for smaller wineries that sell on allocation. It is a powerful external brand-building agent. It also enables wineries to keep raising their prices – again evidence of a Veblen good in action, since demand often rises following price increases.

Is there or was there a secondary market for fine Napa wines, like in Bordeaux and Burgundy? Clearly, auctions have validated such an approach, and several notable Napa wineries are featured benchmarks in such auctions. Several winery owners commented that collectors are an important part of their market – people who buy wine as an investment instrument (perhaps the ultimate liquid asset, as Frank recently argued[23]), possibly to drink at some stage, but also as a hedge in a broad investment portfolio. Such actions affirm the quality staying power of certain Napa brands, transforming wine into investment-grade commodities whose worth continues to be driven by their exclusivity and perceived luxury status.

Exactly how many buy to subsequently sell at a later date is difficult to ascertain. Such hedging behavior by customers certainly captures the essentialism of a luxury product as well as its hedonic ambiguity. On the one hand, gaining access to a fine wine and possessing it – whether for display or the anticipation of future consumption – denotes the audacity of ownership. On the other hand, if its value appreciates considerably in future years it demonstrates the owner's financial acumen. Either way, it is a privileged asset that remarkably synthesizes cultural and monetary capital. And it is at this discerning set of wealthy individuals that many wineries target their product.

In their recent book on luxury-wine marketing, Peter Yeung and Liz Thach argue that there are six essential features that characterize a luxury wine and demarcate it from ultra-premium products. These are: exceptionally high levels of quality, a sense of special place or location, scarcity, high price points, the privilege that consumption confers on the buyer, and the overall sense of pleasure that such a buyer derives from ownership, drinking, and savoring the wine.[24] This certainly characterizes a subset of Napa wineries, such as several of those listed above. Most are typically small producers, selling on allocation at a high price. They have simultaneously shaped the evolving market for wine in the valley as well as benefiting from its emerging reputation. While setting prices very high and limiting availability they have signaled the quality of their product. When subsequently endorsed by critics, they have been able to carve out a segment in the market that caters to a small percentage of the population. But they are not entirely separate from the overall market since other wineries have realized that there is a benefit to premium pricing strategies for a quality product. When such wineries also garnered favorable critic scores, they too have been able to raise prices. Given the growing demand for ultra-premium wine (as well as that in the luxury category), many have capitalized on this trend and seen a boost in sales. All of this has come as wineries have embraced the technical skills and viticultural innovations that foster better and more consistent quality.

Is Cooperation Now Redundant?

I have argued that cooperation and the informal exchange of tacit knowledge were powerful complements to the growth of scientific techniques that were disseminated from university specialists and winegrowing programs. This had been a crucial component of the reputation building for the area. This transparency cemented social relationships that eventually became embedded in an evolving network of winery owners and winemakers. This in turn encouraged trust and cooperation, which shaped the structure of the market in which individual agency became subsumed under an informal collective organizational umbrella. This effectively lowered barriers to entry, since information costs were minimal even if capital costs were rising. However, as more newcomers entered the industry, especially those with strategies based on very limited production and restrictive availability, were these new entrants less likely to embrace such a model?

The so-called cult wineries frequently leveraged their extensive financial resources to contract for the sort of knowledge and techniques that were hitherto part of an embedded collective organizational-learning framework. That and the widespread availability of information crucial to quality production rendered informal sharing less normative. Conceivably, according to some this led to more atomistic behavior, even among winemakers.[25] With much higher stakes along with more extensive incumbent knowledge, there might be disincentives to share detailed knowledge with other winemakers. In fact, attempts to differentiate their product, albeit within the general stylistic format of Napa wine, gave winemakers an opportunity to emphasize the distinctiveness of single estates and even "blocks" within a vineyard. Micromanaging production in ways that brought out the unique flavors of the terrain was seen as indispensable for the creation of a unique wine.

By now, most winemakers had extensive experience (together with formal training), so many felt they no longer needed to engage in the informal networks that sustained their earlier counterparts. Several commented that by 2000 most of the best sites had been planted, and yields were of sufficient quality (and quantity) that there was little value-added knowledge to be derived from informal information exchange. "I always heard that there was a real community of winemakers in the valley – and there is, but it's obvious that some think of themselves as being exclusive and are far less likely to share stuff with others," commented one winemaker who had been employed by three wineries in the past. He went on to say: "Don't get me wrong, as there are some of the cult folks who are down to earth and happy to give tips on lots of things, from trellising and hedging, to fermentation issues. And there are those who you rarely see, and when you do they seem distant and unfriendly."[26]

What had started as an informal cooperative structure appeared to be evolving into a hierarchical network. One winemaker who had been in the industry for twenty-two years, primarily with one winery, stated:

> I still call up other winemakers that I think might be dealing with similar issues to me and see what they're doing, we still taste each other's wines and many of us get together informally. We've been around for many years and we know each other well. But there's a distance now between some of us. Maybe we've just grown big and there's too many folks doing this. But I think that some of the recent boutique wineries have tried really hard to be exclusive and that's thrown up barriers. It's hard to pinpoint, but I just don't feel that there's the community that there was a decade ago. There are some folks who really are not approachable and seem to like it that way. But then again we're so much more competent than we were ten or twenty years ago.[27]

Another winemaker with twenty years' experience in various parts of the world commented how people in Napa are more cautious now with what they say: "Everything's more formal, with various technical groups for winemakers and vineyard managers, which is great ... but it seems more contrived."[28] Much of the earlier transparency appears to have dissipated and individuals appear more cautious and even intent on protecting what they feel might be proprietary information. Winemakers intimated that they still sought advice, but the responses were less spontaneous and more formalized. Several claimed that the formal meetings always seemed to have an element of caution and the spirit of openness had gone. It was as if information had become a resource that could be shared but was very much conditional on one's network position and status.[29] In other words, a subset of elite winemakers had become demarcated from the broader group.

This nascent form of intra-network status differentiation tended to discourage interactions that in the past were knowledge conduits. Arguably, this is because certain wineries who had relied on their Napa affiliation for brand building were now seeking ways of differentiating their position. This meant further emphasis on site specificity, estate designations, and even basic viticultural methods to proffer an aura of exclusivity. As wineries sought an identity derived from the unique circumstances of place (terroir), the skills of their winemaker, and the advice of a famous consultant alongside high scores from critics, they could implicitly separate themselves from others with the valley. This does not mean they denigrated the Napa link – far from it. It merely exemplifies how the distinctive resource bundles they had developed rendered such contact less essential. Furthermore, many of the new cult wineries eschewed the mutually supportive and embedded networks as they pursued more focused niche strategies. One winemaker summarized this changing attitude as follows: "It's all about what we can get out of individual blocks and whether that satisfies Parker or Jim Laube (*Wine Spectator*'s Napa expert). Once that's been figured out there's less reason to get into the details with others, and if I did I'm not sure I'd trust them anymore."[30]

Others were similarly less sanguine about the continued benefits of cooperation. Winemakers will always seek individual solutions to problems they encounter, relying on an increasing body of formal knowledge as well as past practices. But that was traditionally tempered by free-flowing information exchange among peers in what was a relatively non-hierarchical setting (although some clearly had greater status than others). Perhaps now there are fewer "surprise" issues to confront. Extant knowledge provides more prescriptive analytic frameworks, so there is less need for informal channels. Notwithstanding such changes, how much does this altered framework notionally affect the efficiency of innovation within the overall group?

An emerging elite subset of wineries has significantly shaped the market for their wine, using price and scores to signal their credibility (and credentials). Others not in such an august group have nonetheless been able to leverage their own status and position by invoking the AVA affiliation to articulate the efficacy of their wine. While Napa became a niche market within the overall generalist market for wine in America, a further subdivision emerged within that niche. The bureaucratization and hierarchy of knowledge transfer were perhaps inevitable products of increased density and heightened competition, especially between newcomers and incumbents, and the increased professionalism and skillsets of winemakers. The tacit knowledge brokering that had occurred earlier was replaced by more circumscribed behavior conditional on hierarchical settings. Networks were more diffuse and structured, but perhaps less reliant on the exigencies of informal relations. The sector's dynamism had perhaps reached a stage where growth could continue organically and more atomistically, but with more formal governance mechanisms to provide oversight and brand protection. Maintenance of Napa's brand was important since it focused squarely on quality. As far as many were concerned, finally Napa was no longer a place for a mass-produced, low-cost commodity wine. That might have been the past, but was conclusively rejected in the present. And yet, Frank Franzia raised the specter of such a repeat performance as he continued his forays into Napa.

When is a Napa Wine not a Napa Wine?

The answer to the above question, prior to a 2006 US Supreme Court rejection of an appeal from a lower court, was somewhat vague. As the reputation of wines from Napa grew throughout the twentieth century, more and more wineries there sought to protect the integrity of their evolving brand by specifying that grapes under the Napa label should primarily be grown in the valley. But part of the problem was that much of the regulatory framework on geographical origins was predicated on implied notions of identity. A 1986 ruling by the Bureau of Alcohol, Firearms and Tobacco (BAFT) grandfathered wineries that had the name Napa (or one of the appellations), and permitted them to continue using grapes that did not primarily come from Napa (if they had done so). Others were required to have a majority of their grapes come from Napa if the region was so indicated on the label.

Fred Franzia, whom we came across in the previous chapter, had extensive vineyards in the central valley, from which most of his grapes were sourced. In 1993 he bought the Napa Creek brand (a small winery on the Silverado Trail), and the following year he added Rutherford Vintners. And then in 1999 he started construction of a large warehouse, wine-storage, and bottling plant on land be had acquired on the outskirts of the town of Napa.[31] Finally, he added the Napa Ridge brand to his growing portfolio when he bought it from Beringer. As an operation, Beringer was unambiguously seen as a Napa brand, but it had for many years, especially as it expanded production, used grapes grown outside of Napa. As an established player that sold quality as well as cheaper mass-produced wines, Beringer did not elicit much criticism for its established business model. It also had the added gravitas of history. This was not the case when Franzia made no pretense about making a high-quality expensive wine. Even the least-astute observer could see what Franzia's intentions were. He had a legitimate Napa brand, for which he could source large quantities of inexpensive grapes from his operations outside of Napa and be perfectly legal in doing so. He could then use these grapes to vinify a wine that had Napa on the label. In fact, in one of his well-known quips in defense if his practices, he stated: "no one assumes Hawaiian Punch comes from Hawaii."[32]

Hawaiian Punch was relatively inexpensive, and clearly a commodity. Napa wineries, on the other hand, had spent decades assiduously arguing that their wine was precisely not a commodity. It was in their eyes a premium product with a valuable brand identity that was predicated upon quality. Protecting that brand meant eliminating imposters, whose poorer quality and cheap wine could only undermine their credibility. Franzia's duplicity in the whole affair needed to be suppressed.

Once again, one sees the power of organizational bodies representing community interests coming into action. In 2000, the NVVA successfully lobbied the California State Legislature to require any wine with a Napa (or sub-appellation) designation to be made from at least seventy-five percent Napa grapes. Franzia challenged that ruling, and a California appellate court agreed with his claims of violation of his property rights. The question that was banded around referred to longstanding federal statutes designed to prevent consumers being misled. In this case, if Napa was on the label the presumption was that the wine and grapes came from that area. In 2002 a three-judge panel rejected the California state attorney general's defense of the statute that wine must be made primarily from Napa grapes (and presumably that people should not always take things on face value!). With this ruling, Franzia appeared to have emerged triumphant, and set about marketing his famous "two buck chuck" Charles Shaw brand that sold for $1.99. The wine was made in his various facilities in Napa but came from grapes in the central valley, which were considerably cheaper than their Napa counterparts.

By offering the cache of Napa "on the cheap," he was seen as delivering a good wine to the mass public. The question would always be whether his intentions at subterfuge were really that blatant. Was he merely taking legitimate advantage of loopholes in the regulatory framework that had been left loose to satisfy earlier incumbents? When the California Supreme Court reversed the appellate court's earlier ruling in 2004, it did so by arguing that consumers had a right to expect a certain level of quality implied by a label, and that Franzia was deliberately misleading them. Undeterred, he appealed the ruling to the US Supreme Court but that was rejected in 2006. The final verdict reaffirmed the seventy-five percent rule. The NVVA were pleased because they had successfully defended the integrity of the Napa brand. But by then Franzia was shipping over a million cases of Charles Shaw to Trader Joe's stores throughout the country, and was presumably satisfied with the revenue stream that ensued. The wine's label indicates the varietal and the only mention of Napa is on the back, in small print, showing where it was bottled.

It is important to note that Charles Shaw was and is not a bad wine – on the contrary it is recognized as quite good. The problem is that it was a cheap wine, and as long as there might be a presumption that it came from Napa it could only damage a brand that was increasingly assertive in pronouncing its overall excellent quality. Napa wineries had long abandoned any pretense of competing on the basis of price. To do so would be counterintuitive since production costs were rising and margins could only improve if pricing increased. The market that Napa wineries had so carefully nurtured and shaped was predicated on carefully managing supply and demand and constructing appropriate status markers that would confer increasing credibility on their wine. This enabled them to keep raising prices and make wine in a style that was immediately accessible to discerning consumers and applauded by critics. The likes of Fred Franzia were clearly persona non grata in this overall equation.

Are Vineyards Good for the Environment?

We saw earlier how Napa's growth after the 1970s increasingly attracted people whose interest was not in making wine but owning property in the valley. They liked the presumed bucolic charm of life amid the vineyards and built large homes there. However, the Agricultural Preserve Ordinances slowed much of this suburban sprawl, but did not prevent the construction of properties on very large lots that nonetheless complied with the minimum-size requirements. There were fewer new homes, but they tended to be much larger, often on the hillsides overlooking the valley.

By the 1990s the attention was shifting toward the environmental impact of vineyards themselves. While they can appear to be the paragon of agricultural charm, vineyards are frequently treated with pesticides and sprays to eliminate the non-humans who also enjoy setting up residency. The resulting runoffs can be detrimental to water quality, especially that of the Napa River that runs through the valley into the San Pedro Bay in the south. That river became somewhat renowned for its pollution, seen in much of its ecological decrepitude when it flowed through the town of Napa. Although a few wineries formally developed organic practices, many were equivocal about introducing them. They were fearful of the damage caused by pests, which appeared to show continued interest in vines. However, pressure was mounting on addressing these environmental problems and developing more-sustainable practices.

In 2001, the Wine Institute established the Sustainable Wine Growing Program in an attempt to improve educational outreach, focusing on ecological issues, reducing pesticide use, and dealing with soil erosion.[33] It was designed as a bully pit that would shame the major offenders as well as provide some detailed information on what could be done without adversely affecting grape quality or even vineyard productivity. Reluctant at first, more wineries eventually adopted organic practices, and most have adopted sustainable practices that frequently include more crop cover between vineyard rows.[34] Fortunately, buying into the new narrative did not apparently have adverse viticultural effects, at least initially. In fact, it appears to have been a prudent policy that neutralized some of the criticism of wineries' environmental neglect. Trout and other fish have also returned to the Napa river – an aquatic testament to the success of such policies.

The other contentious issue involved increased planting on hillsides and the resulting soil erosion and deforestation. With increased demand for vineyard sites and the lack of further available land on the valley floor, it was inevitable that the hillsides would become sites for new vineyard planting. This gained traction when advocates of hillside planting argued that grape quality could be better there. The thinner soil made vines work harder, so it has been argued, and the result can be a better quality, and a more intensely flavored grape. The fact that some of the best wines were now coming from hillside vineyards lent further credibility to such claims. The growth of hillside plantings resulted in an almost inevitable outcry by some that such practices were damaging the environment. The negative sentiment and opposition gained traction by invoking narratives about "outsiders with too much money" taking over the valley, increasingly making it a playground for the rich.[35] The hillsides were the latest volley in this campaign to slow development. Admittedly, trees and brush were cleared, with the result that heavy rains sometimes washed down soil and silted up the rivers. Hillsides were also more difficult to farm, especially when steep, and questions were raised about practices that might further damage the overall ecology. Such sentiments had been raised from time to time before when permits were somewhat controversially awarded, but the recent opposition was fueled by the growing density of wineries.

Regulatory frameworks have been enhanced by the Board of Supervisors, who struggle to balance the interests of long-time residents, winery owners (new and old), and independent vineyard owners. But it is clear that actions to delimit further expansion have become more rigorous, and presumably will continue to be so. Recent discussions of the Winery Definition Ordinance and whether changes need to be made regarding wineries that were grandfathered in when the original ordinance was passed in 1990 have bought further attention to winery expansion. In 2013, for example, the Napa County Planning Commission debated the ordinance, and eventually agreed that, despite its imperfections, it should stand. One of the contentious issues involved large expansion proposals by two wineries (Reata and Raymond) and whether they would continue to abide by the original seventy-five percent Napa fruit requirement. It was argued by Andy Beckstoffer that enforcement should become stricter since there was no guarantee that wineries could expand production with current Napa grapes, but then use non-Napa grapes for their pre-1990 production.[36] The eventual decision to leave things largely as they are was the least contentious way of resolving the controversy.

The above issues are emblematic of growing pains in an industry that was rapidly becoming a victim of its own success. Groups had become more effectively mobilized within the institutional frameworks that in the past had been developed to provide improved governance and transparency. The problem now was that the groups had multiplied and found a more strident voice, frequently in opposition to further changes that might conceivably disadvantage incumbents. On the other hand, as we have seen, many of the newcomers brought resources that further improved efficiency and enhanced legitimacy for Napa wine. They refined the market in ways that were beneficial to themselves, but in doing so simultaneously angered many old timers and invigorated the overall collective reputation. It was truly a Faustian bargain that some recognized, albeit reluctantly, while others continued to demur.

Consolidation and Growth

Much has been written about the trials and tribulations of the Mondavi family and the winery that still bears their name.[37] Purportedly hurt by years of mismanagement and family squabble, the winery went public in 1993, although the Mondavi family retained ownership of the majority of stock. Ostensibly seen as a way to raise working capital without compromising the Mondavi family members' continued leadership role, the sale did little to alleviate the apparent operational problems. From being one of the pioneers in taking Napa upmarket with the introduction of Opus One plus the continued commitment to improving quality in the Robert Mondavi brand, they nonetheless maintained a significant mass-market presence with their Woodbridge brand, which by the late 1990s was producing just under five million cases a year (much of the grapes coming from Lodi). That and other acquisitions outside of Napa provided them with valuable cash flows to sustain their overall operating costs. But the nagging question was whether they were an ultra-premium winery or mainly a commodity producer. Lacking a clear focus was not the least of their problems as financial troubles mounted. When influential critics questioned the quality of their premium-wine vintages in the early 2000s, blame was directed at winemaker

Tim Mondavi. "Increasingly light, indifferent, innocuous wines," wrote Robert Parker about the reds, "that goes against what Mother Nature had given California."[38] In other words, Mondavi appeared to eschew precisely the style that people had come to expect of Napa Cabernet Sauvignon, and for which they were willing to pay a price premium. Rather than address these escalating problems with their top wines, the company tried to reposition and rebrand their lower-priced wines by raising the prices. But they were immediately confronted by a glut of cheap wine from central-valley competitors. A double storm of quality concerns with the best wines and their loss of low-cost leadership at the bottom made all their wines harder to sell. The result was a financial crisis with profits dropping sixty percent by 2003, and a glut of wine on hand. In response, the company started very public discussions about ways to restructure that included selling off the "luxury" segment, writing down much of its bulk-wine inventory, implementing layoffs, and the sale of vineyard land. However, such scenarios proved unnecessary when in October 2004 it was announced that Constellation Brands would buy the entire Robert Mondavi Corporation for $1.03 billion in cash and $325 million in assumed debt – in excess of the $749–939 million in equity value that RMC executives claimed they could achieve through restructuring.[39] This was an interesting shift in focus for Constellation, whose reputation was for the acquisition of low-end wines. However, their recent purchase of Franciscan Oakville Estates and Mt. Veeder indicated their growing interest in the ultra-premium segment. RMC would effectively enable them to straddle both segments of the market.

Other corporate acquisitions in the first decade of the twenty-first century included Diageo's purchase of stock of the Chalone Wine Group, Duckhorn's sale to private-equity firm GI Partners in 2007, and Gallo's purchase of William Hill Estates in that same year. Earlier, Gallo affirmed their interest in acquiring properties in Napa when they purchased the Martini winery in 2002, suggesting that, like RMC, they wanted to straddle both low-end wines and premium properties. Finally, one of Napa's great surviving pioneers and winner at the famed Paris Tasting, Warren Winiarski, sold Stag's Leap Wine Cellars to a partnership of Ste. Michelle Wine Estates and the Italian wine family consortium Antinori.

None of these sales were the product of hostile takeovers so redolent in the corporate world. Quite the contrary, as many of the above wineries had been shopping for buyers in order to cash out of their business. Some had succession problems, which can often affect generational transitions, but the sentiment of others was cogently expressed by one owner who had earlier sold his estate: "I have always loved Cabernet Sauvignon, so for years successfully made it. But it was always an expensive proposition, so I decided to sell. With the proceeds I bought a large vineyard in the central valley where I sell table grapes to Costco. And now I am able to drink all the great Napa cabs that in the past I couldn't afford!"[40] It's almost inevitable that a cycle of such sales will continue, the normal fluctuations in business activity in a market that remains robust, but also with players who decide that they have had enough and it's time to sit back and enjoy life.

On a smaller scale in the acquisition stakes, two relative newcomers – Pat Roney and Leslie Rudd – bought the Girard winery in 2000 in what was the beginning of Vintage Wine Estates (VWE), which was formally created in 2007. The company eventually acquired over thirty individual wineries, with a core business strategy of buying well-performing brands at different price points[41] and retaining their individual name, but achieving scale economies by shifting all of the back-office activities to a central location in an industrial park in Santa Rosa. It is currently a quarter-billion-dollar business whose Napa brands include Clos Pegase, Cosentino, Delectus, and Swanson Vineyards, as well as Girard. Their other brands are from regions in California as well as Washington State, but their premium ones remain Napa focused. This exemplifies another business model that is gaining traction in Napa – smaller brands brought together under a broader umbrella. Simultaneously pursuing different price points, VWE adheres to a core strategy that seeks to differentiate products for different markets without negating the essentialism of individual winery identity (and, in the case of Napa, their brand integrity).

While corporations were demonstrating a financially motivated interest in winery acquisitions, wealthy individuals were also given a chance to participate in the ownership dream, albeit in a partial way. In 2000, Bill Harlan, owner of Harlan Estate, founded Napa Valley Reserve. Designed as a private winegrowing estate where members, who pay a $165,000 deposit plus annual fees for case production,[42] can participate as much as they like in the making of a wine that they can deem "their own." Currently at approximately six hundred members, the reserve is on an eighty-acre estate that caters to those who like wine and want to experience working in the vineyard and help make their own wine without the encumbrance of vineyard ownership. It is hands-on agricultural labor, but also has a definitive social ambiance, replete with an extensive wine library and full-time librarian. Members cannot sell the wine but can give it to friends or donate it to charity. Ultimately, they are given the chance to make a wine they otherwise would not be able to make, even if they owned vineyards or had the time to devote winery ownership.

As a paean to exclusivity it is perhaps without parallel, and is another example of how Bill Harlan identified a core group of wine lovers who want more than a vicarious experience of Napa's excellence. The Reserve wines are Cabernet based but with other varietals blended that come from outside the estate. Those with the wealth to avail themselves of such an opportunity are also savvy enough to recognize the product of their (limited) involvement is a nonetheless excellent wine. Absent of any positive revenue stream from their endeavors, members presumably derive immense satisfaction from being part of an exclusive enclave. Among likeminded and well-off peers, for a few weeks a year they can live the Napa dream without the inevitable headaches of ownership.

Conclusion

In evaluating Napa's recent successful decades, Charles Sullivan aptly notes, "The idea that one can make fine wine, and lots of wine, but not lots of fine wine, has probably been laid to rest."[43] Napa has patently proved that, with the right resources, appropriate techniques, and a steady supply of consumers who are willing to pay a price premium for a quality wine, even if it is American made, one can be a successful winery. By making quality central to production goals, and by assiduously paying attention to the smallest details, more and more Napa wineries have been able to consistently produce wines that can be easily benchmarked against the best of the world. Admittedly, some of the impetus and drive came from a handful of boutique or cult wineries who set the bar high in terms of the price/quality ratio. But in doing this they contributed to an organizational framework both shaping collective capabilities and creating a viable market for their product. Capitalizing on extensive financial resources that enabled them to build their brand as a quintessential luxury product with no expenses spared, they were also keenly aware of a consumer market that was becoming more knowledgeable about wine as well as having the wealth to purchase it, no matter the price.

Various scholars have argued that California wineries benefitted from classification schemes that prioritize grape varietal designation and a horizontal appellation structure.[44] In other words, the use of certain varietals such as Cabernet Sauvignon has enabled them to charge a higher price for their wine. Furthermore, price becomes the sine qua non signaler of quality, and when combined with high external-critic scores is a credible reputational marker that reinforces pricing power. While wineries have focused primarily on quality control, technical efficiency, and brand identity, they have also gained institutional support by advocating a coherent appellation system that reinforces identity. However, unlike the vertically structured regulatory framework that exists in France, Napa has constructed a more fluid classification that provides organizational constraints but not the external restrictions found in France.

The informal structures that facilitated collective organizational learning from the 1970s to 1990s have gradually been replaced by a more structured dissemination of knowledge. As more winemakers obtained professional training they were better equipped to deal with the various vineyard exigencies, and thus less reliant on network structures. More entrepreneurs entering the industry often had requisite resources that enabled them to contract for essential services, and as transactions became more formalized and systemized, so the dynamic capabilities became atomistic rather than reliant on collective structures. This resulted in enhanced competition between wineries, although the expanded marketspace for consumers seems to have minimized any negative impact on sales. By creating a market for fine wine among consumers becoming more knowledgeable and thus gaining validation from critics, Napa has been able to continue its reputational growth and credibility. In fact, the reliance on objective scores by experts has been a determining feature of many wineries' strategic trajectory.

As networks have evolved and Napa became recognizably the site for great wine, it is not surprising that hierarchical fragmentation occurred. Size and ownership differences alongside increased density have rendered earlier cooperative patterns less formidable and even manageable. That and enhanced governance designed to protect the nascent brand have formalized many operational procedures. Institutional frameworks further circumscribed firm behavior without necessarily constraining differentiation. Community engagement is now more likely to be seen through the lens of privileged access, with incumbent firms and newcomers charting a course that is designed to solidify the reputational gains through more individual agency. Events of the last decade, including fires, environmental issues, labor shortages, land-use patterns, recession, and Covid-19, have brought further strains to the area. On top of that, questions have been raised as to who will successfully replace the baby boomers who were the primary market for expensive wines. In the next and last chapter we will discuss these issues.

Notes

[1] Charles L. Sullivan, *Napa Wine* (San Francisco: Wine Appreciation Guild, 2008), 379.

[2] Frances Scott Morton and Joel M. Podolny, "Love or Money? The Effects of Owner Motivation in the California Wine Industry," *Journal of Industrial Economics* 1, no. 4 (2002): 431–56.

[3] Edith Penrose, *The Theory of the Growth of the Firm* (Oxford: Basil Blackwell, 1959).

[4] See Ian M. Taplin, "Network Structure and Knowledge Transfer in Cluster Evolution," *International Journal of Organizational Analysis* 19, no. 2 (2011): 128.

[5] Sullivan, *Napa Wine*, 386.

[6] Ibid., 385.

[7] Ibid., 388.

[8] In the mid-1990s the average sugar/alcohol level for Cabernets was 23.3/13.5 percent. By the early 2000s it was 25.2/14.6 percent. In other words, these wines became more alcoholic. See Sullivan, *Napa Wine*, 391 for further details.

[9] He states that "unpressed, fermented must, held in sealed tank for weeks after primary fermentation has ended. The heavy, bitter tannins are drawn off, leaving a rich, highly colored, well flavored wine, with noticeably softer tannins." Sullivan, *Napa Wine*, 390.

[10] If a bottle is sold at retail for fifty dollars, then the price paid to the grower is five thousand dollars.

[11] W. B. Gray, "Oakville's Screaming Eagle Winery Sold," *San Francisco Chronicle* (March 23, 2006); James Laube and A. Weed, "A Shake Up at Screaming Eagle," *Wine Spectator* (April 9, 2009).

[12] Personal interview, 2006.

[13] Personal interview, 2005.

[14] Personal interview, 2010

[15] For a longer discussion of this issue as it pertains to Bordeaux, see Stephane Ouvrard and Ian M. Taplin, "Trading in Fine Wine: Institutionalized Efficiency in the Place de Bordeaux System," *Global Business and Organizational Excellence* 37, no. 5 (2018): 14–20.

[16] Personal interview, 2015.

[17] Personal interview, 2006.

[18] Personal interview, 2010.

[19] Personal interview, 2006.

[20] Personal interview, 2016

[21] Personal interview, 2007.

[22] One such General Manager of a winery that was accorded this status told me that "the number of distributors who considered him one of their best friends multiplied five-fold in the days after it was announced that one of his wines had won 'Wine of the Year' award." Personal interview, 2005.

[23] R. T. Frank, "A Cellar Full of Collateral, By the Bottle of By the Case," *New York Times* (July 26, 2015), 3.

[24] Peter Yeung, and Liz Thach, *Luxury Wine Marketing* (London: Infinity Ideas Limited, 2019), 10–11.
[25] See Taplin, "Network Structure and Knowledge Transfer," 138.
[26] Personal interview, 2005.
[27] Personal interview, 2007
[28] Personal interview, 2016.
[29] Taplin, "Network Structure and Knowledge Transfer," 37.
[30] Personal interview, 2006.
[31] Sullivan, *Napa Wine*, 402.
[32] Ibid., 402.
[33] See Wine Institute, "California Wine Community Sustainability Report," 2004.
[34] Sullivan, *Napa Wine*, 407.
[35] One of the most trenchant criticisms of the forces transforming the valley can be found in James Conaway, *The Other Side of Eden* (New York: Houghton Mifflin, 2002). Conaway lays bare what he sees as the intrusion of outsiders awash in cash setting up wineries often as vanity projects with little regard for established operations or other locals.
[36] P. Jensen, "County Planners Stick by Winery Definition Ordinance," *Napa Valley Register* (February 21, 2013).
[37] See for example, Julia F. Siler, *The House of Mondavi* (New York: Gotham Books, 2007).
[38] Robert Parker, *Wine Advocate* (December 2000), 8.
[39] C. Emert, "Legendary California Winery Company is Sold," *SFGate* (November 4, 2004).
[40] Personal interview, 2013.
[41] See B Freeman, "Wine Business Profile: Pat Roney of Vintage Wine Estates," *Forbes* (February 19, 2020).
[42] Members must pay a two thousand dollars annual fee and purchase at least seventy-two bottles at a cost of seventy-five dollars a bottle. Members can produce a maximum of three barrels (nine hundred bottles) a wine annually. They are also given a chance to purchase other Harlan wines that are sold on allocation.
[43] Sullivan, *Napa Wine*, 424.
[44] See W. Wei Zhao, "Social Categories, Classification Systems, and Determinants of Wine Price in the California and French Wine Industries," *Sociological Perspectives* 51, no. 1 (2008): 163–99; "Understanding Classifications: Empirical Evidence from the American and French Wine Industries," *Poetics* 33 (2005); 179–200; Anand Swaminathan, "Resource Partitioning and the Evolution of Specialist Organizations," *Academy of Management Journal* 44 (2001): 1169–85.

CHAPTER SEVEN

CONCLUSION: THE ODYSSEY ENDS

We've come a long way since Hamilton Walker Crabb purchased 240 acres of farmland from George Yount in 1868, planted vines, and founded Hermosa Vineyards. A mixture of slopes and valley floor that included a variety of soil types, eventually this would become one of the best vineyard sites in Napa. When he bought an additional adjacent 119 acres from Eliza Yount in 1881 he paid what was then a fairly large sum of $100 per acre. We'll never know how much Crabb really appreciated the land he bought, or if his purchase was blind luck or an astute assessment of the soil's potential. The fact that he was willing (and able) to pay that sum suggests a certain confidence in what the property would yield. In 1886 he changed the name of his winery from Hermosa to the To Kalon Wine Company and went on to produce large quantities of wine with different grape varietals from this and other vineyard sites he had acquired. Alas, phylloxera outbreaks in the 1890s devastated his vines and he died penniless in 1899. Subsequent To Kalon owners (including Mondavi and Opus One, using grapes from there, and now Andy Beckstoffer) have been much more fortunate. They realized the full potential of this site and produced some of Napa's most famous wines. That this site is something special, from which one can make a wine that is truly exceptional, is perhaps the ultimate affirmation of Crabb's early assessment.

When Crabb died there were almost two hundred wineries in Napa, and after a shaky start in the late nineteenth century, by the early 1900s the quality of wine was beginning to gain recognition in some international tastings. Despite a wide range of varietals, Cabernet-based ones were even then seen as making some of the most notable wine. Then Prohibition came, and the industry largely collapsed and would not be resuscitated until the 1960s. When growth resumed it was initially slow, but by the 1980s, according to the Alcohol Beverage Control of California, there was an average of three new wineries registered each year. By the 1990s it had risen to seven. During this time, many failed or were taken over by corporations. But those that survived provided impetus to newcomers, many of whom brought extensive resources that enabled them to thrive.

This slow growth pattern has changed dramatically in recent decades. Since 2015 there has been an average of 170 new wineries registered each year. Currently, there are seven hundred grape growers in Napa, and 475 physical wineries. While the latter are the public face of the valley's wine industry, the former are testament to the continued fascination that people have with starting a winery, however small and limited the production. I will return to a discussion of the small wineries shortly. However, it is important to note that small remains the norm. Approximately eighty percent of Napa wineries have less than ten thousand annual case production, and seventy percent have less than five thousand, while ninety-two percent are family owned.

As noted in the previous chapter, the recent rapid growth has not come without problems: local residents bemoan the large number of tourists; there is an inadequate supply of low-income housing for workers; and winery owners complain about a shortage of vineyard labor and how expensive it has become, as well as the mounting tide of regulations as local government attempts to address many of the emerging environmental concerns. Some say there are too many wineries now, but how many wineries can be considered too many? Do new wineries nonetheless bring valuable assets that bestow collective efficiency gains, or are they merely additions to the growing density? Have new permit rules become a barrier to entry to all but the really wealthy? And does the 1990 Winery Definition Ordinance, reaffirmed in 2008 to extend protection to 2058, and which prevents tasting rooms for many newly established wineries, unfairly discriminate against small family-run operations? Does code compliance present almost insurmountable barriers for small wineries? Are extant land-use practices damaging the ecology of the valley? On top of this litany of concerns and complaints have been the droughts and fires of recent years, plus the earthquake in 2014 that caused an estimated $1 billion in damage. And then in 2020 Covid-19 hit!

These are daunting concerns, and yet, when the industry was in its infancy, people were struggling to determine what to grow, how to grow it, and ultimately how it could be sold. Technical innovations and science eventually solved many of these issues. But do the new events present problems that might curtail some of the recent success and sustained patterns of growth?

Success Begets Fame and Controversy

Somewhat Phoenix-like, Napa appears to have a resilience that defies the odds and has emerged, not necessarily unscathed, as one of the world's premier wine regions. Napa is not unique in accomplishing this success, but the speed with which it has achieved it is quite remarkable. A world-class wine industry has been created within the past few decades, as opposed to centuries in the Old World. Admittedly, Napa has the perfect climate for growing Bordeaux varietals, and winegrowers have been the recipients of much accumulated knowledge, from viticultural techniques to disease mitigation, which their earlier counterparts in other countries conspicuously lacked. Its success is testament to generations of trial and error, the gradual application of scientific techniques, extensive financial resources by newcomers, and an increasingly discerning consumer who appreciates fine wine and is prepared to pay a premium for it. True to neoclassical dictates, a balanced supply-and-demand equilibrium has facilitated growth. But as I have argued throughout this book, it was the way in which these variables came into play, creating a dynamism that shaped market formation and fostering a network structure, that was conditional upon cooperation. Success came though systematic attempts to organize the market in ways that recognized the primacy of offering a premium and ultra-premium product, and doing so by managing production, pricing, and availability in ways that satisfied the status aspirations of many consumers.

Napa's recent growth was more than a mere tale of agglomeration economies as the driving force of innovation or even institutionalized categories of status classifications that have led to the region's reputation. Instead, one has seen a gradual and purposeful attempt, often by key individual actors, to develop and shape the structure of wine production around quality parameters while sustaining a collective organizational-learning framework. In the past, they might not always have been successful and, as we have seen, some failed in these endeavors, only to have the torch taken by a subsequent new entrant to the industry. But, in recent decades the successes have outweighed the failures.

What has thrived in this innovative environment has been a willingness to seek and adopt technical skills marked by a heightened scientific approach to winemaking. Experimentation and innovation were promulgated. Without eliding traditional practices (the recognition of beneficial site-specific characteristics that have made Cabernet Sauvignon the signature grape is salient here), the largely collective commitment toward embracing quality over quantity, despite earlier reservations and opposition, was solidified in the decades after the 1960s. The perspicacity of systematized and technically informed viticulture was crucial in overcoming earlier failures, and eventually permitted winemakers to develop stylistic characteristics that would differentiate Napa from other regions' wines. Key individual actors played a singular role, but so did legions of professionally trained winemakers able to forcefully advance the cause of scientific and technical rigor. And they did this for decades through informal information-sharing networks. Their training lent legitimacy and credibility to the often-innovative approaches that became formalized in Napa (and California in general). At the same time, an evolving institutional structure providing requisite governance to substantiate organizational and production norms was supported and encouraged by many winegrowers. They saw this as a way to enhance and then protect the brand that they had worked so assiduously to establish. It became a protective umbrella that fostered efficiency and credibility.

The final piece in the journey toward legitimacy came from powerful industry critics whose often high praise for certain wines proved to be the ultimate oenological stamp of approval. Their role minimized information asymmetries for consumers since their scores could affirm the reputational signals of price and brand that wineries promulgated. In turn, this reassured consumers who might otherwise feel insecure purchasing an expensive wine. More confident and wealthy consumers subsequently translated into increased sales.

The unique way in which a market for fine wine has been developed in Napa is testament to the creative dynamism of the above actors. They have simultaneously contributed to an emerging wine culture in the United States and shaped its formation. Largely eschewing price-based competitive strategies, most wineries in Napa have adopted a clear differentiation approach based on developing a distinctive individual and collective brand identity. They have capitalized on an increasingly informed customer base whose discretionary incomes permit indulgences of ephemeral products such as wine. And some have done this through an unambiguous appeal to status markers that the wealthy cherish. Such wines are expensive, difficult to buy, and acknowledged to be of exceptional quality. These are wines at the luxury end of what is an ultra-premium market. While having a small subset in terms of volume, such wineries have had a disproportionate effect in their ordering and organization of that market. Their reputational accolades have percolated down to other producers, for whom the residual benefits of brand identity have accrued.

Arguably, most businesses are run by rational utility maximizers. What has made Napa somewhat unique in the past few decades is not the absence of such behavior but the way in which it has become bifurcated. There are winery owners for whom lifestyle and status comprised the driving force behind their utility, thus obviating the need for profit, at least in the early years. In such cases, ownership of a winery was the ultimate expression of success. In that same camp, but with a slightly different goal, were those whose dream was to build a winery that would have a place among the pillars of global oenological excellence. No expense was spared in developing this vision. Both deferred positive revenue streams for much longer than would be viable for most businesses (with the possible current exception of some new technology companies that embrace growth at the expense of profits).

For others, however, more immediate profit maximization was a necessity. With different timeframes dependent on resources, eventually they needed positive cash flows. This did not necessarily mean sacrificing commitment to quality – they simply averred that profits should eventually accompany such a goal sooner rather than later. Nonetheless, they shared the commitment to excellence that the first group embraced – they differed in the financial parameters within which that goal could be reached. Whether they were bullied, via benchmarking, into accepting practices that newcomers from the first group introduced is not always clear. But to remain competitive often necessitated approximating the new norms, and many of these norms were being rewritten by newcomers. To do otherwise was potentially problematic. Eventually, however, the operating procedures of both groups coalesced around accepted if not always identical patterns. Any significant differences now are likely to be size related – for instance, small luxury wineries committing to labor-intensive practices more than large ones.

These differing definitions of rational utility have coexisted in Napa in recent decades, becoming more pronounced as many of the latest newcomers clearly embraced the utility maximizing stance. Separate and hardly equal, the differing types have nonetheless come together to present, if not a unified face, then at least one that suggests some consensus. Importantly, both have shaped the industry's evolution with their interactions, further fostering the growth of dynamic capabilities.

Having argued that cooperative network interactions were crucial to the industry's growth, I show how a more hierarchical network structure appears to have emerged in the last two decades, partly conditioned by the differential resource attributions that characterized the two groups. This has led to increased competition as density has also developed – with more wineries in the same region, using the same differentiation strategy, competing for the same customer base. Almost inevitably, this has led to a diminution of the cooperative culture that sustained the valley's incipient development. It has also contributed to some of the criticisms directed at wealthy winery owners who are seen to be indifferent to the needs and interests of other small family-run wineries that are less generously financed.[1] Be that as it may, it does not appear to have adversely affected the region's growing legitimacy. In fact, the contrary has possibly occurred, as media reports bring attention to not the disaffected but those who have engineered success through high-profile activities.

One such seminal event that is emblematic of the latter is Auction Napa Valley. Founded in 1981 by several prominent winegrowers (including Robert and Margit Mondavi) and sponsored by the Napa Valley Vintners Association, it was based on a similar charity event in Burgundy, the Hospices-de-Beaune Wine Auction, founded in the fifteenth century. As with the latter, Napa wineries donate wine, and in most years more than one hundred lots can be sampled and eventually auctioned with the proceeds going to twenty-five local nonprofits and strategic initiatives in community health and children's education.[2] The auction has proved extremely popular over the decades, with sustained interest from most key wineries. One can only speculate that there is normative pressure for top wineries to participate, hence its ongoing popularity.

Aside from its charitable benefits, it nonetheless further credentials Napa wineries by virtue of de facto association with an iconic Burgundian event. It invokes the same aura of charitable respectability as that venerable auction – the willingness of key actors to give back to the community in which they are located. But participation also signals the identity of key players in the prestigious wine market – an unofficial barometer for excellence, and thus reputation building. Participants are acknowledged for their beneficence but also recognized as being part of the industry elite. Napa's initiation of such an event augments winery status and is yet more evidence (without intending to evince cynicism) of how excellence has been organized in a strategic and purposeful manner.

Production and Consumption Trends

Cabernet Sauvignon continues to be the signature grape for Napa, with acreage doubling from just under ten thousand acres in 1995 to almost twenty thousand in 2017. Meanwhile, Chardonnay, Merlot, Syrah, and Pinot

Noir acreage has remained about the same. Median grape prices for all varietals have increased from $1,100 per ton in 1992 to $5,900 in 2017. Looking specifically at Cabernet Sauvignon, the average price was $7,509 per ton that year, a continued upward trend from previous years. The highest price paid was $50,000 per ton and the lowest $1,000. This compares with Merlot ($3,387 per ton), Zinfandel ($3,623), Syrah ($3,783), and Pinot Noir ($2,798). For whites, the highest average was for Chardonnay at $2,809. Compared with neighboring Sonoma County, where red and white varietal pricing is approximately the same, the exceptions are Merlot, where it is much less at $1,918, and most notably Cabernet Sauvignon, where it is $3,073 – less than half the price paid in Napa for that grape. The 2020 figures on varietal percentages by value indicate Cabernet Sauvignon at fifty-five percent, Chardonnay at fourteen percent, Merlot at ten percent, Pinot Noir at six percent, Sauvignon Blanc at six percent, and finally Zinfandel (the valley's old workhorse) at three percent. Napa clearly conveys pricing power to Cabernet Sauvignon growers.

In recent years, a series of good harvests has led to an increased supply of grapes, putting pressure on pricing. This has been particularly problematic for bulk-wine producers, but has raised some questions over continued premiumization strategies and whether prices can continue to be pushed upward. However, the silver lining from the fires of 2020 was that yields were dramatically lower in Napa, and this has reduced the oversupply of grapes there. According to a recent Silicon Valley Bank report, thirty-five percent of wineries reported record low yields, and thirty-eight percent weaker than average yields.[3] Since some wineries have decided to not issue a vintage for that year because of vineyard damage or smoke taint, this will eventually smooth out any supply-and-demand imbalances.

In years when vintage quantity has been high, some top wineries have sold wine on the bulk market rather than increase their own production. Others have sold grapes that don't quite meet their specifications to other wineries. Screaming Eagle winemaker Nick Gislason commented on this practice at a MW event in Napa when he noted that, of the fifty blocks their grapes are sourced from, only a few grapes make it into the two wines he makes – the remainder are sold under strict nondisclosure agreements.[4] This practice is not uncommon and is an important source of quality grapes for small production wineries, many of whom do not own vineyards. Under such a system, their wine is then marketed by online wine sellers such as *Wine Access*. As part of a growing e-commerce trend, companies such as this perform what is essentially an arbitrage role, securing small lots of wine from incipient brands that lack a bricks-and-mortar presence (a virtual winery), but can nonetheless claim unique provenance for their wine via their vineyard source. It is a valuable means of moving small quantities of wine over the internet to consumers who are knowledgeable yet rely on the curatorial role of e-commerce company experts, most of whom have professional wine training (i.e. as a sommelier or, in some cases, master of wine credential). Such digital sites have also become a valuable platform for established brands who find that they are an effective way of moving some of their wine without recourse to normal distribution channels (and presumably saving some of the associated markups).

Another solution to excess supply is the development of a second label. Grapes that do not quite meet the standards for a premium wine, or simply be from insufficiently mature vines for the exacting standards of a main wine, can be processed and sold under a second or even third label. One such winery owner told me that he saw this as a way of offering a wine from younger vines at an approachable price ($140 a bottle), designed to appeal to millennials and hopefully future consumers of his much more expensive wines. There's always the risk that this could dilute the overall brand, but since his wine is all sold on allocation there seemed little chance of this happening.

The idea of a second label is well established in France, where it is viewed as part of an "affordable luxury" strategy.[5] It is designed to give people an entry-level luxury product that might satisfy their status seeking at a more modest price. By giving access to the brand for those who might not otherwise have the financial means, builds a potential future relationship. Knowledgeable consumers might be aware of such differentiation and seek such wines because they are still perceived as being of high quality but more affordable. However, the latter term is relative since the price points are generally still much higher than average, and thus appealing to a small subset of the population. Some Napa owners indicate the success of such a strategy in helping them build their brand with younger consumers as typically the latter cannot even get on allocation lists. Others appeared adamantly opposed to this approach, with one saying that when he tentatively tried such an idea he was unable to get his distributors to take the wine and promote it.[6] He went on to further add that he was concerned about devaluing his brand, and since then has preferred to sell excess capacity on the bulk market.

Over the past few decades the wine market in general has seen a shift toward premiumization (wines priced over twenty dollars a bottle). Almost seventy percent of wine sold by volume is under nine dollars a bottle, but that generic category continues to decline as consumers trade up. In the nine to nineteen dollars range (affordable premium) there is some growth in sales, but not at the level one might have expected as new consumer groups enter the market. In the past, younger groups would adopt this category as they gradually aged, and this was the pattern among the baby-boomer generation. However, younger consumers are far less brand loyal, are attracted to spirits and even craft beer, and are hobbled with far more debt than their counterparts forty years ago. Recent reports on consumption trends indicate that millennials do not appear to be adopting the per-capita wine consumption practices that were anticipated.[7] When combined with the continued aging of baby boomers – the formidable market for luxury wines – this much anticipated generational transition has stalled. The mutability of millennials whose lack of discretionary income often impairs excess spending has also slowed the growth of per-capita wine consumption in the United States since 2016. Baby boomers with seventy percent of US discretionary income and half of the country's net worth are gradually declining in numbers and scaling down their purchasing habits.[8] In the luxury category where they have dominated they are augmented somewhat by high net-worth groups from finance, technology, and the sports industry. Whether this latter group can be a satisfactory replacement for the boomers and if younger consumers eventually step up remain to be seen.

These demographic trends compound the oversupply of grapes following abundant harvests in recent years, although this has enabled some wineries to focus on further improvements in quality and charging accordingly. In fact, general price increases work best for luxury wines rather than the overall super-premium category (twenty-five dollars and up), since demand continues to exceed supply at the very high end of the market. For wineries not in that refined space, the challenge is finding the right path to consumers. Many wineries, especially smaller ones, have refined their focus on relationship building with customers. This can include the use of brand managers, social media, and digital marketing as well as hospitality to leverage an enduring connection between winery and customers.[9] It is also a crucial part of sales-channels innovations that have changed in the past ten to fifteen years, and which have provided significant new retail opportunities.

The Rise of Direct-to-consumer Sales

Ever since the US Supreme Court ruling in the Granholm versus Heald case in 2005,[10] permitting out-of-state direct sales (and shipments) to consumers, small wineries have seen this as a way to build a consumer base and bypass normal distribution channels (and the associated markups). Since then, there has been a surge in direct-to-consumer (DTC) sales as small wineries use their tasting rooms to build up a customer base, either through a wine club or internet marketing. In 2000, small wineries sold a mere fifteen percent of their wines directly to consumers. Now that figure is closer to sixty percent (almost evenly split between the tasting room and the wine club). Being able to sell directly to consumers has been one reason behind the growth of wineries in the past two decades, notwithstanding rising startup costs since use permits and code compliance can sometimes cost up to half a million dollars.

Further compounding these changes in the channel landscape has been consolidation trends among distributors. Prior to the 2005 US Supreme Court ruling, most wineries relied upon the three-tiered distribution system or sold direct from their winery. Since the late 1990s there has been considerable consolidation and concentration in the distribution industry, which left many small wineries bereft of a viable distributor.[11] For example, in 1995 there were three thousand distributors, whereas by 2017 that number had fallen to twelve thousand.[12] Large distributors were less likely to give shelf space to small firms simply because the latter lacked the volume to reach multiple markets. Typically, distributors look for strong brands with predictable sales, which tends to exclude smaller wineries, especially startups. Even though newer boutique-style distributors were spawned after this consolidation and served niche markets, they often proved insufficient for the needs of smaller wineries who shifted to the DTC model. Several small wineries indicated that DTC sales were perhaps fifty percent of their volume but up to seventy-five percent of their revenue. Avoiding the middleman and the multiple markups improved their margins, but often necessitated additional staff to deal with individual state shipping requirement and licenses. Factoring in these ancillary costs negates some of the benefits of DTC sales channels, and yet the latter clearly offers a wide range of benefits, especially if additional customer information can be gleaned from such sales and used to further future transactions.

Clearly, many small wineries that have managed to become established have benefitted from Napa as a destination site, especially among those high-income millennials who seek experiential luxury.[13] Luxurious spas, exclusive restaurants, and concierge-type winery visits are attractive to such a demographic. Wineries have responded by creating intimate and personalized experiences, providing a level of hospitality that is the consummate expression of uniqueness, and that differentiates the winery from others. "Make the visit memorable and long-lasting," commented one winery owner. "We want to make people feel special, to give them a chance to relax, try our wine but also have an intimate setting that is well scripted to make them feel at ease and not be intimidated."[14]

Notwithstanding concerns about excessive numbers of tourists, wineries have leveraged this group to establish more personal relationships and develop a consumer base. Given the large number of tourists intent on visiting wineries, most Napa wineries now charge high tasting fees (often seventy-five dollars and up) and limit numbers by appointment. When justifying the high tasting fee, one owner of a 2,200-case winery said that it limited:

> the numbers to those who actually can afford to buy the wine. When it was lower we had far more people come and taste, but a much smaller percentage actually bought more than a bottle, and few joined the wine club. Now it's much more manageable, with fewer people we can spend more time and develop a relationship. They tend to be repeat customers.[15]

One owner of a small winery commented:

> We make 1,300 cases and we don't have lots of dotcom money to throw around, so there's no way we can realistically compete with the cult of even corporate owned wineries. We can make good wine and look for smaller distributors to sell it, rely upon a few restaurants to promote it, and the rest goes through our wine club. Really, we don't compete in the same marketplace as the others, yet we share a similar product.[16]

When tasting rooms reopened in late spring 2020 after the Covid-19 shutdown in March, they did so with much-limited capacities, and often by appointment only. This resulted in far fewer visitors, but enabled wineries to develop a more elevated experience for those that came. Together with more personalized attention and a longer time to talk with customers and regale them with appropriate stories about the winery and the wine, wineries have been able to foster relationships that large numbers in the past inhibited. Many of the top wineries had always done this, seeing it as a way to nurture a select customer base for repeat sales. Now, many more have followed suit and found that customer numbers are often halved but sales remain the same as pre-Covid-19. Fewer people are apparently buying more wine. It has also given wineries the opportunity to collect more systematic data on their customers that can be used for future direct sales. However, the majority of wineries still lack sophisticated data-analytic skills and staff, so managing such data can be difficult.[17] This could prove problematic going forward. Interestingly, wineries that sell only on allocation have been able to collect detailed customer information, yet, somewhat ironically, they need it less for future sales potential.

In the early spring of 2020, when tasting rooms were shut, wineries were forced to pivot away from that sales channel to e-commerce, particularly online sales, even to people not in their wine club. At that same time a massive channel shift occurred with off-premise sales as demand for wine in grocery stores often led to depleted supplies. This proved challenging for wineries that did not have an established brand or the resources to manage e-commerce. However, intermediary digital-sales channels such as Wine Access and Wine.com provided some respite. Given demographic trends and younger consumers' willingness and enthusiasm for using e-commerce, it is perhaps incumbent for wineries to develop this capability even if tasting rooms return to their previous role. To not do so could jeopardize future sales potential, especially if the current customer base of boomers slows their wine purchasing.

Also, a more customized approach might be needed to reach consumers who might not travel to wineries but still want to experience the pleasure that associational links can provide. Whether this is through virtual tastings, brand ambassadors, or being made to feel special through more informal communications, each demonstrate important relational capabilities that can leverage sales. Again, digital skills are increasingly part of the new dynamic capabilities that will allow wineries to benefit from channel changes. Developing such capabilities and turning them into core competences is the current requisite phase of organizational excellence.

Summary of Different Types

What is the current pattern of ownership structure of wineries in Napa? A proximate breakdown reveals the following general categories. Seventy percent are small family-owned wineries, producing less than five thousand cases per year, who sell eighty percent of their wine DTC with the remaining going to select restaurants or through a targeted distribution system to certain cities and states. The next category (also often but not exclusively family owned) comprises wineries with higher production volumes, often in the thirty-five to one hundred thousand-case level. They sell DTC, but mainly rely on wide distribution networks, nationally and in some cases internationally. They have a strong brand presence with sales in restaurants, resorts, and even cruise lines. Thirdly are wineries that have been acquired by corporate bodies, through mergers and acquisitions (M&A), or even by other large wineries, retaining their brand name but managed by a broad corporate organization. Some such wineries often retain their small image and original brand identity; others are larger, and have a national presence. These wineries rely on professional and experienced management teams, and, as noted earlier, gain scale efficiencies by consolidating back-office functions to a central location. In the past decade there have been many sales of family-owned wineries, including those in the top categories. For example, in 2007 a controlling interest in Duckhorn Vineyards (est. 1976 by Dan and Margaret Duckhorn) was acquired by a private-equity firm, GI Partners, who in turn sold it to another private-equity firm, TSG Consumer Partners, in 2016. Such sales are emblematic of returning corporate interest in Napa wineries and their revenue potential. On a grander scale, but nonetheless reflective of changing ownership trends, E & J Gallo in December 2020 finally secured agreement from the Federal Trade Commission to acquire more than thirty wine brands from Constellation Brands for $810 million. Most of these brands are mass market in the US and overseas, but this has led some to question whether it is the new corporatization in wine, leading to increased power over many aspects of the industry by a few key players.[18] Even though this acquisition has minimal impact on most Napa producers at the higher end of the market, it nonetheless speaks of the growing concentration of ownership structure. For instance, Gallo now has over thirty percent market share in the low-price segment, and it does still make some ultra-premium wine in Napa, so it is not entirely absent from the valley.

Finally are the luxury brands or cult wineries which rely on sales through allocation systems. Their production levels are low (generally not exceeding two to three thousand cases annually), with availability extremely limited. But they too have experienced new ownership acquisitions in recent years, from the case of Stan Kroenke's purchase of Screaming Eagle (referred to in an earlier chapter) to luxury conglomerate LVMH's purchase of a sixty percent stake in Colgin Cellars (est. 1992) in 2017 that added to their other Napa majority acquisition of Newton Vineyard (est. 1977) in 2001. And in 2019 Grace Family Vineyards was sold to a San Francisco management consultant. In each case, brand identity was retained, and often extant production staff continue to operate in their original roles. But in many cases the owner is a corporate body, and not a family.

Can Small Wineries be Economically Viable?

In this changing landscape, where scale and capital resources are ever more important, small family-owned wineries might find themselves increasingly at a disadvantage. One option that has rendered them more viable has been the use of custom-crush facilities. Some families with a longstanding presence in the valley as vineyard owners who sold grapes to wineries have realized the value-added potential of developing their own label. The capital costs of starting a winery for small volume production can be prohibitive, so they have turned to one of the numerous custom-crush facilities that have a long tradition in Napa. In earlier chapters I alluded to the several cooperative wineries that existed earlier in the twentieth century which essentially functioned in a similar way to a contemporary custom crush. However, current custom-crush facilities operate on a purely contractual basis, where wine is made for independent growers, whereas the earlier cooperatives were a way of small operations joining together to pool their resources.

Making wine in custom-crush facilities such as the Napa Wine Company,[19] obviating the need for fixed production-facility costs and even tasting rooms, also eliminates many of the regulatory costs of a winery startup. Such facilities provide extensive services, from crushing, fermenting, blending, barrel storage, packaging, and compliance. Essentially, this gives a vineyard owner – or even someone who might not even own land and merely purchases grapes – the opportunity to develop a brand while retaining a small production volume without the significant overheads of an actual winery. Basically, one pays for whatever services required, presumably calculating the charges against the fixed costs of establishing one's own production facility. It does provide wineries with asset-light flexibility, and has been a crucial part of small brand development in recent decades. This category of wineries can exploit digital-sales channels to enhance their position and develop new sets of dynamic capabilities that could enable them to reach new younger consumers, especially if their price points are competitive.

Since the majority of wineries are small, how are they faring in this changing competitive environment? It is difficult to precisely determine gross margins, although generally the ultra-premium wineries do better than others. In terms of actual production costs, when small wineries use custom crush they outsource the professional services of a winemaker and focus their attention on marketing and sales. This can enable them to capture capital equipment-cost savings. If they are too small to generate consistent critic evaluation they rely on informal networks and relationship building with key constituencies to sell their wine. Those that are most successful in this task have either found boutique distributors to provide limited wholesale distribution to key markets, or rely on DTC for the majority of their sales.

There are size disadvantages as well as benefits simultaneously. It is easier to sell under five thousand cases directly from the tasting room or through a wine club than it is ten thousand. One winery owner told me that she abandoned her goal of being a ten thousand-case winery and reduced production to four thousand cases.[20] Ninety percent of that is now sold DTC, and all on allocation. On the other hand, staffing costs cannot be easily spread over a higher-volume production where scale economies come into play. A thirty to forty thousand-case winery has the benefit of scale, a smaller production-team ratio to sales, and the opportunity to garner critic scores and placement nationally in high-end restaurants. Brand positioning (and value) for such a winery relies on pricing strategies that position them comparably to peer competitors, consistent quality, the ability to leverage site specificity to further signify status, and exogenous forces in the broader market (for example, economic conditions and political exigencies).

With the possible exception of the very top luxury wineries, the challenge most face is the gradual and potential further decline in their core customer base of baby boomers, whose large discretionary income has sustained growth since the 1990s. There is uncertainty whether millennials and generation Xers will take up the wine baton. Neither group currently has the discretionary income of boomers. Millennials have shown interest in alcoholic beverages but included craft beer and cocktails in their repertoire. They have also been most likely to embrace healthy lifestyles, which can mean less alcohol consumption, and are far less brand loyal than previous generations. Added to these issues has been the rise of neo-prohibition views in recent years, with wine consumption viewed as possibly deleterious to ones' health.[21] Consumers have apparently taken notice as per-capita wine consumption growth has slowed in recent years. At the moment, much information on this is anecdotal, although such concerns possibly resonate among millennials more than other generations.

Given that Napa wineries have spent decades cementing their brand, organizing a market around it, and capitalizing on lifestyle imagery that invokes conviviality in wine consumption, any of the above changes can be potentially problematic. However, millennials' continued interest in experiential luxury can nonetheless resonate with winery visits in what is an undoubtedly elegant and sophisticated environment. In fact, Napa wineries are particularly well positioned to capitalize on precisely this form of luxury behavior by wealthier millennials (as well as the other demographic groups). Careful curation of the winery experience with a focus on exclusivity can have the same latent status appeal to such groups as for their older, wealthier counterparts. Napa has become a destination site built solidly around the wine experience, and been explicit about fomenting a wine culture conditional on winery visitation. Unlike Bordeaux and Burgundy, where winery visits are not the operational norm, Napa has thrived by making access positively ceremonial. As wineries move from a tasting-room model, where customers stand at the bar and drink through the various wines on offer, to one that is more choreographed and exclusive, often paired with food, it consolidates their ultra-premium reputation and focuses on consumers who are able to purchase the wines.

Current Problems

In addition to the aforementioned demographic imponderables, environmental and climatic issues have been particularly problematic in recent years. Severe droughts and high summer temperatures have forced wineries to rethink their vineyard maintenance and production practices. Fires over the past two years have caused extensive property damage to a number of wineries, with production facilities destroyed and storage sites impaired. However, the pressing concern from the 2020 fires has been smoke taint, and wineries are currently investigating the extent to which this might have occurred. Some have already written down that year's harvest, and several have indicated they would not make wine from this vintage. All have said they would rebuild, and they presumably have the financial resources to withstand this temporary loss. But the question on many minds is whether this is the new normal, and if so, what can be done to mitigate the potential damage? The other pressing issue is whether the same cooperative spirit that sustained early developments can be resuscitated to develop a systematic community response and preparation for catastrophic events. Arguably, such cohesion could induce a more proactive approach, as opposed to ex post facto damage control.

Other environmental concerns mentioned in the previous chapter, however, do appear to have been more cogently addressed. The Agricultural Preserve initiatives are seen as being particularly successful. According to Napa Valley Vintners, ninety percent of Napa County is under permanent or at least a high level of protection from development, and at the end of 2020 it was estimated that the green environmental certification program was almost ninety-five percent complete.[22]

Another set of concerns stems from changes in immigration law and the availability and price of farm labor. Recent curtailment of immigration, both illegal and legal, has contributed to a shortage of seasonal labor. This has been an ongoing problem for decades, but has become more accentuated in the past few years. House prices in the valley have skyrocketed, leaving laborers with long journey times from cheaper areas outside the valley. Attempts have been made to provide temporary housing for workers, such as the three publicly owned and operated farmworker centers that each accommodate sixty workers and are open nine to eleven months a year.[23] This has proved helpful, but also insufficient for the high demand. All of this has been further compounded by the growth of cannabis as a legal crop, which has attracted would-be vineyard workers with higher pay and less-arduous working conditions. In fact, many winery owners cite cannabis as a potential serious problem that will further contribute to labor shortages.

One obvious solution to the above labor shortages lies in automation. Technological innovations, such as automated picking and sorting, have become viable and used by some vineyards. However, that runs counter to the artisanal aura of handpicking and sorting that is invoked by top wineries, and has been publicly rejected by many. It remains to be seen whether such resistance will persist in the face of continued labor problems and the apparent utility of automation. Somewhat ironically, it seems that certain practices can legitimately embrace technical efficiency, while others are deemed incompatible with the notional quality mandates. Technical rigor jumpstarted Napa in the latter part of the twentieth century, but perhaps it didn't question the essentialism of human-grape interactions!

Concluding Comments

Napa's journey to becoming a world-class wine region has been relatively short, but quite tortuous. I have argued that, once the potential for producing high-quality and even exceptional wines was recognized, and individuals with requisite resources entered the industry willing to pursue production strategies commensurate with that goal, Napa's position was firmly established. As consumers displayed a willingness to transfer their allegiances from Old World to Napa wines, and influential critics rendered information asymmetries less redolent, demand conditions changed in ways that were clearly favorable for this new class of producers. By organizing and shaping an evolving marketplace for their product, and by carefully nurturing brand development, winegrowers transformed what had hitherto been a commodity product into an ultra-premium one, and even in some cases a luxury good. Earlier patterns of cooperation and informal information sharing perhaps inevitably dissipated as technical skills were more easily and formally disseminated. Also, as winery density increased, the spillover benefits of such exchanges were less normative as competition replaced cooperation. This is not to say that collaboration is redundant – far from it, as can be seen from the formal associational groups that advocate for the collective image of Napa (and protect its valuable brand). It is just less salient and normative. Perhaps this is the inevitable evolution of an industry that, having gained international credibility, has also become the embodiment of a formalized and hierarchical structure. Governance structures have become embedded in institutional frameworks that simultaneously constrain actors as well as permitting heightened agency on the part of the more powerful players. But owners' ability to organize the industry remains powerful and less constrained by regulatory frameworks than their Old World counterparts, yet increasingly cognizant of environmental, political, and demographic exigencies. Confidence, albeit in an uncertain future, abounds, and the reputational credentials remain firmly intact. In many respects, they continue to be masters of their own destiny.

Unlike many of its Old World counterparts, Napa comprises a small geographic area, being thirty miles from north to south and only five miles wide. Much of the available land is already farmed, and there are few potential vineyard sites left because the remaining land is either too hilly or protected by zoning rules. This has increasingly posed problems for some wineries that would like to expand production but simply lack acreage, and thus grape supplies. Inevitably, if demand continues to rise then one can only presume that prices will increase.

While accounting for a mere six percent of California grapes harvested, Napa is responsible for twenty-seven percent of sales value. This is a definitive statement on its continued eminent position as a premium wine-producing region. And it is Cabernet Sauvignon that has become the signature grape upon which much of the valley's recognition rests. With their rich texture and copious fruit, Napa Cabernets are seen as stylistically quite distinct. They have become the embodiment of Napa's wine success story, at their best possibly revealing an aesthetic dimension that places them in the rarified world of unique goods. Other varietals contribute to the story, but it is Cabernet that has positioned the area at the pantheon of vinous excellence. To this success one attributes much hard work, extensive resources, a clear market strategy, and consumers whose growing knowledge (and wealth) made them eager participants in the adventure.

Notes

[1] James Conaway, *The Other Side of Eden* (New York: Houghton Mifflin, 2002).
[2] See Auction Napa Valley for further details. https://auctionnapavalley.org/about. Thus far, it has donated over two hundred million dollars to such local charities.
[3] Silicon Valley Bank. 2021. "State of the US Wine Industry 2021," 30, https://www.svb.com/trends-insights/reports/wine-report-2021.pdf.
[4] Peter Yeung and Liz Thatch, *Luxury Wine Marketing* (London: Infinite Ideas Limited, 2019), 87–8.
[5] Linda Lisa Maria Turunen, *Interpretations of Luxury* (Cham, Switzerland: Palgrave, 2018); Jean-Noël Kapferer, *Kapferer on Luxury* (London: Kogan Page, 2015).
[6] Personal interview, 2019.
[7] Silicon Valley Bank, 2020, "State of the US Wine Industry 2020," https://www.svb.com/trends-insights/reports/wine-report-2020.pdf.
[8] Silicon Valley Bank, 2021, 8.
[9] Yeung and Thatch, *Luxury Wine Marketing*, 189.
[10] *Granholm v. Heald*, 544 U. S. 460 (2005), United States Supreme Court.
[11] Ian M. Taplin, "Competitive Pressures and Strategic Repositioning in the California Premium Wine Industry," *International Journal of Wine Marketing* 18, no. 1 (2006): 61–70.
[12] Andrew Adams, "The Challenge of Distributor Consolidation," *Wines and Vines* (September 2017).
[13] Ian M. Taplin, *The Evolution of Luxury* (London: Routledge/Taylor Francis, 2020), chapter seven; Joanne Roberts and John Armitage, *The Third Realm of Luxury* (London: Bloomsbury, 2020).
[14] Personal interview, 2019.
[15] Personal interview, 2019.
[16] Personal interview, 2009.
[17] Silicon Valley Bank, 2021, 25.
[18] Emily Saladino, "Gallo, Constellation and the New Corporatization in Wine," *Wine Enthusiast* (January 21, 2021).
[19] The earliest iteration of the Napa Wine Company was in 1877 when two Frenchman bought land and planted grapes to create the ninth bonded winery in California. It has subsequently been owned by Heublein, who made Chardonnay under the Inglenook label, and then sold to the Pelissa family in 1989. By the mid-1990s it had become a state-of-the-art custom-crush facility. It currently caters to several dozen clients and its production ranges from 250 to 100,000 cases. See https://napawineco.com.
[20] Personal interview, 2016.
[21] In the late 1980s and early 1990s, reports on French Paradox suggested that moderate wine consumption had health benefits, including low rates of cardiovascular disease. However, recent guidelines have challenged these assumptions, arguing for even lower levels of daily consumption for a healthy lifestyle. For a full discussion of this evolving debate see Silicon Valley Bank, 2021, 42–4.

[22] See https://napavintners.com/press/docs/napa_valley_fast_facts.pdf.
[23] See https://www.countyofnapa.org/467/Housing-Authority.

APPENDIX A

WINERY INTERVIEW DETAILS

All of the interviews were conducted in accordance with normal social-science research protocols. Confidentiality was guaranteed with the agreement that any statements interviewees made that were cited in the published work would be non-attributable. Interviews lasted between one and three hours and were conducted on the business premises of the owner. The only three exceptions were when an interview was conducted in a coffee shop, which was the preferred location out of convenience for the interviewee. In every case I explained what my basic questions were, and indicated that this was part of a longstanding research project, the results of which would be used in my lectures and academic publications. Everyone had the option to decline to respond to any question they were uncomfortable with. In the text, attribution is by date (year) to maintain confidentiality.

List of wineries where interviews were conducted and name of the person interviewed plus date

Arger-Martucci	Katerena Arger, Daughter of owner	2005
Arietta	Fritz Hatton, Proprietor	2020
Baldacci Family Vineyards	Debbie Cali	2006
Ballentine Vineyards	Bruce Devlin, Winemaker	2005
Cain Vineyard and Winery	Chris Howell, GM and Winemaker	2006
Cakebread Cellars	Bruce Cakebread, President & CEO	2007
Cardinale	Clay Gregory, President	2006
Chase Family Winery	Pam, Co-owner	2006
Chateau Boswell	Susan Boswell, Owner	2016
Clark–Claudon Vineyards	Laurie Claudon, Co-Owner	2006
Clos Pegase	Samantha Rudd	2016
Colgin Cellars	Manager	2007
Corison	Cathy Corison, Owner and winemaker	2007
Cuvaison Estate Wines	Jay Schuppert, President and CEO	2007
Ehlers Estate	K. Morrisey, Winemaker and GM	2016
Far Niente	Larry McGuire	2005
Frank Family Vineyards	Todd Graff, Winemaker	2006

Girard Winery	Steve Ross, GM	2006
Groth Vineyards Winery	Suzanne Groth	2020
Groth Vineyards Winery	Carl Ebbeson, Controller	2005
Harlan Estate	Bill Harlan, Founder	2015
Harlan Estate	Don Weaver, Estate Director	2006
Hartwell Vineyards	Linda LaPonza, GM	2007
Hourglass Winery	Michael Cooperman, International Sales Manager	2017
John Anthony	John Anthony Truchard	2006
Jones Family Vineyards	Rick Jones, Owner	2016
Joseph Phelps	Tom Shelton, President and CEO	2005
Keever Vineyards	Olga Keever, Co-owner	2019
Kerr Cellars	Curtis Hecker, Director of Sales, GM	2019
Lambourne Family Wines	Mike Lambourne, Owner	2020
Lang and Reed	John Skupny, Proprietor	2007
Milat Estate Winery	Cliff Little	2005
Miner Family Winery	Dave Miner, Owner	2005
Morlet Family Vineyards	Luc Morlet, Owner	2017
Opus One	Dave Pearson, CEO	2006
Palmaz	Christian Palmaz, President	2016
Pahlmeyer	Cleo Pahlmeyer, President	2019
Pahlmeyer	Ed Hogan, GM and Director of Sales and Marketing	2007
Picayune Cellars	Claire Ducrocq, Owner, Winemaker	2017
Plump Jack Winery	John Connor, GM	2005
Quintessa	Charles Thomas, Winemaker	2010
Reynolds Family Winery	Steve Reynolds	2006
Rudd Winery	Charles Thomas, Vice President Winemaker	2005
Shafer	Doug Shafer, Owner	2005
Sherwin Family Vineyards	Steve Sherwin, Owner	2006
Signorello Vineyards	Chris Carmichael, National Sales Director	2005
Silver Oak Cellars	David Duncan, GM	2005
Silverado	Russ Weis, President	2005
Spottswoode Estate	Beth Novak Milliken, President	2005
Spring Mountain Vineyard	Jac Cole, Winemaker	2005
Stag's Leap Cellars	Russell Joy, Vice President	2019
Tor	Tor Kenward, Owner	2007
Tres Sabores	Julie Johnson, Owner and winemaker	2007
Viader	Delia Viader, Owner	2016
Vineyard 29	Chuck McMinn, Owner	2006
Vine Cliff Winery	Christine Peterson, Sales and Marketing Manager	2005
Vintage Wine Estates	Pat Roney, President and Co-Founder	2016

Whitehall Lane Winery	Mike McLaughlin, GM	2005
Zahtila Vineyards	Laura Zahtila, President	2007

BIBLIOGRAPHY

Adams, Andrew. 2020. "The Challenge of Distributor Consolidation." *Wines and Vines* (September). https://winesvinesanalytics.com/sections/printout_article.cfm?article=feature&content=189049.

Adams, Leon D. 1973. *The Wines of America*. Boston: Houghton Mifflin.

Amerine, Maynard, and Albert J. Winkler. 1944. "Composition and Quality of Musts and Wines of California Grapes." *Hilgardia* 15, no. 6.

Aspers, Patrik. 2011. *Markets*. Cambridge, Polity Press.

Auction Napa Valley. https://auctionnapavalley.org/about.

Bazin, Jean-François. 2013. *Histoire du vin le Bourgogne*. Paris: Ēditions Jean-Paul Gisserot.

Beckstoffer, W. Andy. n. d. "Premium California Vineyardist, Entrepreneur, 1960s to 2000s." Regional Oral History Office, Bancroft Library, Berkeley, University of California. https://www.lib.berkeley.edu/libraries/bancroft-library.

Best, Michael. 2018. *How Growth Really Happens*. Princeton: Princeton University Press.

Blazer, Robert L. 1948. *California's Best Wines*. Los Angeles: Ward Richie Press.

Brands, Hal, and Charles Edel. 2019. *The Lessons of Tragedy*. New Haven: Yale University Press.

Breschi, Stefano and Lissoni, Francesco. 2001. "Knowledge Spillovers and Local Innovation Systems: A Critical Survey." *Industrial and Corporate Change* 10, no. 4: 975–1005.

Briscoe, John. 2018. *Crush: The Triumph of California Wine*. Reno: University of Nevada Press.

Brook, Stephen. 2011. *The Finest Wines of California*. Berkeley: University of California Press.

Carroll, Glenn. 1985. "Concentration and Specialization: Dynamics of Niche Width in Populations of Organizations." *American Journal of Sociology* 90: 1262–83.

Carosso, Vincent P. 1951. *The California Wine Industry, 1830–1895*. Berkeley: University of California Press.

Chapuis, Claude, and Steve Charters. 2014. "The World of Wine." In *Wine Business Management*, edited by Steve Charters and Jérôme Gallo, 13–23. Paris: Pearson France.

Charters, Steve, and Jérôme Gallo (eds.). 2014. *Wine Business Management*. Paris: Pearson France.

Chauvin, Pierre Marie. 2011. "Architecture des prix et morphologie sociale du marché. Les cases des Grands Crus de Bordeaux." *Revue Française de Sociologie* 52, no. 2: 277–309.

—. 2019. "Globalization and Reputation Dynamics: The Case of Bordeaux Wines." In *The Globalization of Wine*, edited by David Inglis and Anna-Mari Almila, 103–14. London, Bloomsbury.

Chevet, Jean-Michel, Eva Fernandez, Eric Giraud-Héraud, and Vincente Pinilla. 2018. "France." In *Wine Globalization: A New Comparative History*, edited by Kym Anderson and Vicente Pinilla, 55–91. Cambridge: Cambridge University Press.

Conaway, James. 2002. *The Other Side of Eden*. New York: Houghton Mifflin.

Coyle, Diana. 2020. *Markets, State and People*. Princeton: Princeton University Press.

Cruess, William V. 1937. "Suggestions for Increasing Consumer Interest." *Wines and Vines* (November 18). https://winesvinesanalytics.com.

Delacroix, Jacques, and Michael E. Solt, 1989. "Niche Formation and Entrepreneurship in the California Wine Industry 1941–1984." In *Ecological Models of Organizations*, edited by Glenn Carroll, 53–70. Cambridge, MA: Ballinger.

Demossier, Marion. 2018. *Burgundy: The Global Story of Terroir*. New York: Berghahn.

Emert, Carol. 2004. "Legendary California Company is Sold." *SF Gate* (November 4). https://www.sfgate.com/bayarea/article/Legendary-California-wine-company-is-sold-2637611.php.

Fiske, J. 1965. "Napa County." *Wines and Vines* (February). https://winesvinesanalytics.com.

Fligstein, Neil. 2001. *The Architecture of Markets*. Princeton: Princeton University Press.

Frank, Robert T. 2015. "A Cellar Full of Collateral, by the Bottle or by the Case." *New York Times* (July 26). https://www.nytimes.com/2015/07/26/business/a-cellar-full-of-collateral-by-the-bottle-or-the-case.html.

Freeman, Brian. 2020. "Wine Business Profile: Pat Roney of Vintage Wine Estates." *Forbes* (February 19).

https://www.forbes.com/sites/brianfreedman/2020/02/19/wine-business-profile-pat-roney-of-vintage-wine-estates.

Gorny, Ronald. 1996. "Viticulture in Ancient Anatolia." In *The Origins and Ancient History of Wine*, edited by P. E. McGovern, S. J. Fleming, and S. H. Katz. Luxembourg: Gordon and Breach Publishers.

Gray, W. Blake. 2003. "Oakville's Screaming Eagle Winery Sold." *San Francisco Chronicle* (March 23). https://www.sfgate.com/wine/article/Oakville-s-Screaming-Eagle-Winery-sold-2501190.php.

Guthey, Greig T. 2008. "Agro-industrial Conventions: Some Evidence from Northern California's Wine Industry." *The Geographic Journal* 174, no. 2: 138–48.

Heimoff, Steve. 2008. *New Classic Winemakers of California*. Berkeley: University of California Press.

Hiaring, P. E. 1979. "Martha's Vineyard." *Wines and Vines* (May): 70–3, https://winesvinesanalytics.com.

Inglis, David. 2019. "Wine Globalization: Longer Term Dynamics and Contemporary Patterns." In *The Globalization of Wine*, edited by David Inglis and Anna-Mari Almila, 21–46. London: Bloomsbury.

Inglis, David, and Anna-Mari Almila. 2019. *The Globalization of Wine*. London: Bloomsbury.

Jensen, P. 2013. "County Planners Stick by Winery Definition Ordinance." *Napa Valley Register* (February 21). https://napavalleyregister.com/news/local/county-planners-stick-by-winery-definition-ordinance/article_0cc3cc3a-7bce-11e2-a2b7-001a4bcf887a.html.

Kapferer, Jean-Noël. 2015. *Kapferer on Luxury*. London: Kogan Page.

Kramer, Matt. 2004. *New California Wine*. Philadelphia: Running Press Books.

Lapsley, James. 1996. *Bottled Poetry: Napa Winemaking from Prohibition to the Modern Era*. Berkeley: University of California Press.

Laube, James. 1989. *California's Great Cabernets*. San Francisco: Wine Spectator Press.

—. 2001. "The Glory that was Inglenook." *Wine Spectator* (October 17). https://www.winespectator.com/articles/the-glory-that-was-inglenook-1057.

—. 2015. "Master of Style." *Wine Spectator* (November 15): 64.

Laube, James, and Augustus Weed. 2009. "A Shake Up at Screaming Eagle." *Wine Spectator* (April 9). https://www.winespectator.com/articles/a-shake-up-at-screaming-eagle-4724.

Leggett, H. L. 1941. *The Early History of Wine Production in California.* San Francisco: Wine Institute.
Marcus, Kim, 2015. "Bill Harlan's Amazing Saga." *Wine Spectator* (November 15): 50–62.
Markham, Dewey. 1998. *A History of the Bordeaux Classification System.* New York: Wiley.
Marks, Denton. 2015. *Wine and Economics.* Cheltenham: Edward Elgar.
Marshall, Alfred. 1890. *Principles of Economics,* London: MacMillan.
Maskell, Peter. 2001. "Towards a Knowledge-based Theory of the Geographical Cluster." *Industrial and Corporate Change* 10, no. 4: 921–43.
Mathews, John A. 2003. "Competitive Dynamics and Economic Learning: An Extended Resource-based View." *Industrial and Corporate Change* 12, no. 1: 115–45.
Mondavi, Robert. 1998. *Harvests of Joy: My Passion for Excellence.* New York: Harcourt Brace.
Napa County Agricultural Commissioner. Annual Report, 1967 and 1980, https://www.countyofnapa.org/AgCom.
Ourvrard, Stephane, and Ian M. Taplin. 2018. "Trading in Fine Wine: Institutionalized Efficiency in the Place de Bordeaux System." *Global Business and Organizational Excellence* 37, no. 5: 14–20.
—. Remaud, Hervé, and Ian Taplin. 2019. "The Bordeaux Classified Growth System." In *Accounting for Alcohol,* edited by Martin Quinn and João Oliveira, 206–22. London: Routledge/Taylor and Francis.
Parker, Robert. 2000. *Wine Advocate.* https://www.robertparker.com.
Penrose, Edith. 1959. *Theory of the Growth of the Firm.* Oxford: Blackwell Press.
Phillips, Rod. 2016. *French Wine: A History.* Berkeley: University of California Press.
—. 2017. *9000 Years of Wine.* Vancouver: Whitecap Books.
Pinney, Thomas. 1989. *A History of Wine in America; From the Beginnings to Prohibition.* Berkeley: University of California Press.
Pninou, Ernest. P., and Sidney Greenleaf. 1954. *Winemaking in California: III. The California Wine Association.* San Francisco: Porpoise Bookshop.
Ridgeway, Cecilia, 2019. *Status: Why is it Everywhere? Why Does it Matter?* New York: Russell Sage Foundation.
Rinzler, J. W. 2018. *Daring to Stand Alone. An Entrepreneur's Journey.* Petaluma: Cameron + Company.
Roberts, Joanne, and John Armitage. 2020. *The Third Realm of Luxury.* London: Bloomsbury.

Saladino, Emily. 2021. "Gallo, Constellation and the New Corporatization in Wine." *Wine Enthusiast* (January). https://www.winemag.com/2021/01/15/gallo-wine-corporatization.

Scott Morton, Francis, and Joel Podolny. 2002. "Love or Money? The Effects of Owner Motivation in the California Wine Industry." *Journal of Industrial Economics* 1, no. 4: 431–56.

Scruton, Roger. 2009. *I Drink Therefore I Am*. London: Bloomsbury.

Shafer, Doug. 2012. *A Vineyard in Napa*. Berkeley: University of California Press.

Siler, Julia F. 2007. *The House of Mondavi*. New York: Gotham Books.

Silicon Valley Bank. 2020. "State of the US Wine Industry 2020." https://www.svb.com/trends-insights/reports/wine-report-2020.pdf.

Silicon Valley Bank. 2021. "State of the US Wine Industry 2021." https://www.svb.com/trends-insights/reports/wine-report-2021.pdf.

Simpson, James. 2011. *Creating Wine: The Emergence of a World Industry, 1840–1914*. Princeton: Princeton University Press.

Staber, Udo. 2001. "The Structure of Networks in Industrial Districts." *International Journal of Urban and Regional Research* 25, no. 3: 537–52.

Storper, Michael. 1997. *The Regional World: Territorial Development in a Global Economy*. New York: Guilford Press.

Storper, Michael, and Anthony J. Venables. 2004. "Buzz: Face to Face Contact and the Urban Economy." *Journal of Economic Geography* 4, no. 4: 351–70.

Sullivan, Charles L. 2008. *Napa Wine*. San Francisco: Wine Appreciation Guild.

Swaminathan, Anand. 1995. "The Proliferation of Specialist Organizations in the American Wine Industry, 1941–1990." *Administrative Science Quarterly* 40: 653–80.

—. 2001. "Resource Partitioning and the Evolution of Specialist Organizations." *Academy of Management Journal* 44: 1169–85.

Taber, George. 2006. *The Judgment of Paris*. New York: Simon and Schuster.

Taplin, Ian M. 2006. "Competitive Pressures and Strategic Repositioning in the California Premium Wine Industry." *International Journal of Wine Marketing* 18, no. 1: 61–70.

—. 2011. "Network Structure and Knowledge Transfer in Cluster Evolution." *International Journal of Organizational Analysis* 19, no. 2: 127–45.

—. 2011. *The Modern American Wine Industry*. London: Pickering and Chatto.

—. 2020. *The Evolution of Luxury*. London: Routledge/Taylor Francis.
—. 2021. "Narratives of Science and Culture in Wine Making." In *Routledge Handbook of Wine and Culture*, edited by Steve Charters, Marion Demossier, J. Dutton, G. Harding, Jennifer S. Maguire, Denton Marks, and Tim Unwin. London; Routledge/Taylor Francis.
Tchelistscheff Oral History. 1979. Bancroft Library, Regional Oral History Office, 112. https://www.lib.berkeley.edu/libraries/bancroft-library.
Todd, Cain. 2010. *The Philosophy of Wine*. Montreal: McGill-Queen's University Press.
Turunen, Linda Lisa Maria. 2018. *Interpretations of Luxury*. Cham, Switzerland: Palgrave.
University of California College of Agriculture. 1896. *Report of the Viticultural Work during the Seasons 1887–93*. Sacramento.
Unwin, Tim. 1991. *Wine and the Vine: An Historical Geography of Viticulture and the Wine Trade*. London: Routledge.
West, G. Page, and Terry W. Noel. 2009. "The Impact of Knowledge Resources on New Venture Performance." *Journal of Small Business Management* 47, no. 1: 1–22.
Worobiec, MaryAnne. 2020. "Creating Her Own Legacy." *Wine Spectator* (May 31): 40–50.
Wine Institute. 2004. "California Wine Community Sustainability Report." https://www.sustainablewinegrowing.org/aboutcswa.php.
Winkler, Albert J. 1972. "Viticultural Research at the University of California, Davis, 1921–71."
https://www.lib.berkeley.edu/libraries/bancroft-library.
Varriano, John 2010. *Wine: A Cultural History*. London: Reaktion Books.
Yeung, Peter, and Liz Thatch. 2019. *Luxury Wine Marketing*. London: Infinite Ideas Limited.
Zhao, Wei. 2005. "Understanding Classifications: Empirical Evidence from the American and French Wine Industries." *Poetics* 3, no. 3: 179–200.
—. 2008. "Social Categories, Classification Systems, and Determinants of Wine Price in the California and French Wine Industries." *Sociological Perspectives* 51, no. 1: 163–99.
Zhao, Wei, and X. Zhou. 2011. "Status Inconsistency and Product Valuation in the California Wine Market." *Organization Science* 22, no. 6: 1435–48.

INDEX

Note: The abbreviation 'n' next to a page number refers to an endnote. The abbreviation 'illus' refers to maps/drawings.

A
Abbey of Saint-Germaine-des-Près (Paris) 33
abstemious behaviour *see* temperance
acreage (vineyards/wineries) 53, 67, 89, 98, 99, 146, 151
  high-quality wines 120, 129, 130, 135–6
  modern-day trends 176–7, 187
advocacy organizations 73, 113, 157, 163
  pre-/post-Prohibition 60–1, 62, 63, 64, 68
  pre-/post-Second World War networking 80–2, 86–7, 91
"affordable luxury" wines 178
Agricultural Preserve Ordinances 99, 113, 161, 185
Ahern, Abbey 114n4
Ahern, Albert 79
Alcohol Beverage Control (California) 170
Algeria 40
Alicante Bouschet 2, 20, 67, 74, 82, 112, 146
Altamira, Father Jose 46
American Journal of Enology 90
American Society of Enologists (1950) 83
American Viticultural Area (AVA) 106, 132–3
American Viticultural Association (AVA) 20, 158
Amerine, Maynard 73, 77, 83
amphorae (Greek storage vessels) 28

Anatolian region (modern-day Turkey) 27
Ancient Egyptians 7, 24, 25, 27, 30
Ancient Greece 4–5, 7, 8, 17, 24, 25, 27, 28–30
Angelica 52
Antinori (family consortium, Italy) 164
AOC *see* Appellation d'Origine Contrôlée
Apothic Red 2
Appellation d'Origine Contrôlée (AOC) 16, 34, 37, 41, 104, 132, 141n30
Araujo, Bart and Daphne 149
Araujo Estate Wines 149
Association of Five 114n25
Auction Napa Valley 175
automated picking/sorting 186
AVA *see* American Viticultural Area; American Viticultural Association
AxR#1 (phylloxera-resistant rootstock) 117–18, 130–1

B
Bacchus (Roman god of agriculture and wine) 30
Bale, Edward Turner 48
Bank of America 96
banquets (*convivium*, Roman Empire) 30
Barolo (Italy) 1
Barrett, Jim 105
BATF *see* Bureau of Alcohol, Tobacco, and Firearms

Beaulieu Vineyards (winery) 73, 75, 77–80, 87, 88, 89, 92n14, 92n26, 95, 107, 128
Beckstoffer, Andrew 107–8, 112, 128, 134, 163, 171
beer-drinking 7, 27, 32, 65, 100, 178, 184
Benedictine monastic order 33, 46, 68
Beringer (winery) 75, 88, 159
Berrouet, Jean-Claude 124
Best, Michael 14
blended wines 41, 53, 54, 59, 60, 62, 63, 69, 74
Board of State Viticultural Commissioners (1880) 55, 57
Bond (brand) 125
Bonnet, Leon 78
Bordeaux 4, 9, 15, 18, 57–8, 62, 64, 172
   as a benchmark 120, 126, 127, 128, 136, 137, 150
   Harlan Estate 124, 125
   historical background 25, 32, 35-37, 38*illus*, 39–44, 45n20
   Paris Tasting (1976) 20, 102, 103, 109
   see also Cabernet Sauvignon
Boston 49
"boutique" wineries 97, 103, 105, 123, 127, 153, 157, 166, 179, 183–4
brand identity 61, 64, 120, 121, 151, 173, 181, 182
   definition of Napa wineries 132, 133, 134, 159–61
   pre-/post-Second World War 36, 39, 44, 58–9, 67, 68, 73, 76, 79, 81–2, 85–6, 91
   price strategies 9, 15, 184
   reputation building 101–6, 111, 113, 139, 144, 149, 153, 154, 157, 158
Brannan, Sam 52
Bréjoux, Pierre 104
Bronco (brand) 133
Bryant, Don 127

Bryant Family Vineyard 149
Buena Vista (winery) 52
bulk wines *see* mass-produced wines
Bureau of Alcohol, Tobacco, and Firearms (BATF) 106, 133, 141n31, 159
Burger 52
Burgundy 3, 4, 15, 17, 61, 68, 175, 185
   historical background 25, 33, 34, 36, 37, 38–9, 71n39
   Paris Tasting (1976) 102, 103, 105
   reputation building 39, 40, 42, 44, 46
Burgundy, Philip, Duke of 34

C
Cabernet Franc 107, 111, 124, 127, 136, 146, 149
Cabernet Sauvignon 1, 20, 57, 62, 90, 143, 144, 166, 168n8, 170
   Napa's signature grape 122, 131, 148, 149, 164, 172, 175–6, 187
   post-Prohibition 74, 75, 78, 79, 82, 84, 89
   quality wine and 117, 118, 119, 120, 124–8, 130, 135–9, 146–8
   reputation building 94–6, 98, 103, 107, 108, 111, 112
   see also Bordeaux
Cain (winery) 111
California 17, 18, 19, 44, 46–7, 48*illus*, 49–51, 67–8
   emergence of quality wine in Napa Valley 51–7, 58*illus*, 59
   overproduction and economic downturn 59–61
   Prohibition and collapse of wine industry 65–8
   resurgence of Napa Valley 61–5
California Farmer, The 50
California Grape Protective

Association (1908) 66
California Mission 48*illus*
California State Agricultural Society
 (1854) 50, 55
California State Legislature 160
California State Viticultural Society
 (1872) 55
California Wine Association (CWA)
 (1894) 60–1, 62, 63, 64, 68
California Wine-Growers
 Association (1862) 55
California Wine Makers' Corporation
 (CWC) (1894) 60, 61
cannabis farming 186
canopy management practices 107
Carignan 67
Carignane 82, 89
Carneros 77, 79, 88, 97, 108
Carpy, Charles 96, 114n4
Cato 30
Caymus Rancho 48, 127
Chalone Wine Group 164
Champagne region (France) 2, 36,
 40, 41, 97
Chandon Blanc de Noir 97
Chandon Brut 97
Chappellet, Donn 89, 95
chaptalization (added sugar) 54,
 71n19
Charbono 52
Chardonnay 68, 82, 94, 96, 97, 112,
 135, 188n19
 acreage 89, 90, 98, 146, 175, 176
 Paris Tasting (1976) 103, 105
 quality wine 117, 118, 119, 120,
 123, 130
Charles Krug (winery) 52, 54, 55,
 59, 79, 87, 92n26, 95
Charles Shaw (brand) 160–1
Chartrons (wine warehouse district)
 (C19th Bordeaux) 38*illus*
Chateau Boswell (winery) 111
Château Giscours 104
Château Haut-Brion 104, 114n25
Château Haute Brion (*Ho Bryn*)
 (Bordeaux) 9, 104

Château Margaux (Bordeaux) 9
Chateau Montelena (Calistoga) 88,
 95, 104, 105, 113n1
Château Mouton Rothschild
 (winery) 39, 104, 109, 111,
 114n25
Château Pétrus 123
château production model 123
chateâux 9, 38, 39, 43
Chenin Blanc 146
China 42
Christian Brothers 75, 88, 95, 129,
 141n14
Cistercian monastic order 33, 46, 68
Cîteaux (monastery, Burgundy) 33
claret (*clairet*) 35, 45n20
classification systems 9, 15, 18, 27,
 36, 39, 44, 64, 166, 173
 American Viticultural Area
 (AVA) 106, 132–3
 Appellation d'Origine Contrôlée
 (AOC) 16, 34, 37, 41, 104,
 132, 141n30
climatic conditions 16, 17, 77, 106,
 118, 124, 131, 136, 172, 185
 early years 50, 51, 52, 53, 56–7,
 61, 62, 69, 71n39
Clos de Vougeot (Burgundy) 34
Clos Pegase (brand) 165
Coca Cola 101
Colgin Cellars (winery) 149, 182
collective organizational learning
 14, 15, 19, 25, 145, 174, 175
 California 50, 54–5, 61, 62, 73–
 4, 80, 95, 105, 116, 128
 France 41–2
 redundancy of 155–8, 167, 187
 universities and 77, 83, 88, 91,
 138
Columella 30
"Commission upon the Ways and
 Means best adapted to promote
 the Improvement and Growth of
 the Grape-vine in California"
 (1861) 50–1

competition 14–16, 19, 20, 21, 60, 144–5, 167, 175, 187
Conaway, James 169n35
Constellation Brands 164, 182
consumers and consumption patterns 1, 4, 5, 6, 7, 13, 15, 16, 17, 42
    Ancient World 24, 25, 27, 28–30, 31–2
    C16th onwards 35–42
    California in C19th 49, 59
    Middle Ages 5, 8, 25, 32–5
    Prohibition/post-Prohibition and 65–6, 80
    quality wine 128, 129, 136–7, 139–40
    rise of wine culture *xi*, 119–22, 143, 148–52, 173, 177–180, 186–7, 188n21
    Second World War and afterwards 81–2, 83–5, 94, 96, 100–2, 112, 116
    wineries/consumer interface 152–5, 178–81
Corison, Cathy 137–8
corporate ventures and investment 120, 129, 164–5, 180–2
    rise of wine culture and 88–9, 94, 95, 96, 97–8, 101, 106–11
    see also ownership
Cosentino (brand) 165
courtiers see wine brokers
Covid-19 pandemic 167, 171, 180
Coyle, Diane 91
Crabb, Hamilton Walker 170
Crane, George 52
Cruess, W.V. 77
culinary life 1, 3, 4, 6, 11, 24, 28, 116, 147, 152
cult wines 21, 148–52, 153, 156, 157–8, 166, 182
custom-crush facilities 183, 188n19
CWA *see* California Wine Association
CWC *see* California Wine Makers' Corporation

D
Daily Alta California 53
Dalle Valle Vineyards 149
Daniel Jr., John 79, 92n14, 123, 126, 141n19
Decanter 10
De Latour, Fernande 92n14
De Latour, George 62, 64, 73, 78, 80, 114n25
Delectus (brand) 165
Demossier, Marion 69, 71n39
Department of Agriculture (USDA) 62
De Pontac, Arnaud 9
Deuer, George 79
Diageo 164
Dionysus (Greek god of wine and ecstasy) 29
direct-to-consumer sales (DTC) 178–81, 184
diseases *see* phylloxera
distilling industry 63, 65, 66, 79
distribution 21, 25, 30, 61, 63, 82, 95, 169n42
    direct-to-consumer sales (DTC) 178–81, 184
    e-commerce 176–7, 181, 183
    transportation 15, 35, 39, 77
    wine merchants and 49, 54
    wineries/consumer interface 152–5, 178–81
Domaine Chandon 97, 108
Dominus (winery) 123–4
Domitian 31
Dom Perignon vintage Champagne 2
dotcom boom 143, 153, 180
drink-driving laws 42
drip-irrigation systems 107
dry farmed (unirrigated) vineyards 75
DTC *See* direct-to-consumer sales
Duckhorn, Dan and Margaret 182
Duckhorn Vineyards (winery) 164, 182
Duncan, Raymond 95
Dunn, Randy 127, 137–8
Dunn (winery) 111, 127, 137–8

E
e-commerce 176–7, 181, 183
Eighteenth Amendment (prohibition) (1919) 66
Eisele Vineyard 149
elitism 5, 7, 8, 21, 43, 85, 111, 119
  historical background 27, 30, 31, 32, 33, 34, 36
  sales and 150, 153–4, 155
England 25, 35, 37, 45n20
entrepreneurship 1, 14, 22, 94–8, 111, 119, 120–9, 167
environmental issues 21, 99, 127, 161–3, 167, 171, 185, 187
Eshcol (winery, later Trefethen) 57, 58*illus*, 67
estate-bottled wine 58, 79, 81, 85, 86
exclusivity 21, 27, 31, 32, 43, 149–53, 155, 157, 166, 185
extant social-science theory 22
extended maceration 136, 147

F
Far Niente (winery) 57
Federal Alcohol Administration 75
Federal Trade Commission 182
fermentation techniques 27, 56, 57, 100, 120, 126, 136, 147, 156
  technological/scientific innovation 72, 73, 77, 78, 79, 83
Fernandez, Elias 122
Finegan, Robert 103
fires 21, 167, 171, 176, 185
firm clusters 14–15
First World War 41
flavor profiles 10
Flemish traders 35
Fligstein, Neil 13
Forni, Antonio 114n4
Forni, Charles 92n14
Forni Vineyard 88
fortified wines 63, 73, 80, 86, 90
Foster, Mark 114n4

France 7, 8, 9, 15, 18, 25, 72, 85, 110, 167
  historical background 32, 33, 34, 35, 41, 42, 44, 46
  Paris Tasting (1976) 20, 102–6, 109, 111–13, 116, 123, 138, 139, 164
  regulatory framework 16, 34, 37, 41, 104, 132, 141n30
Franciscan missionaries 46–7, 48*illus*, 68
Franciscan Oakville Estates 164
Frank, R.T. 154
Franzia, Frank/Fred 119, 133, 158, 159, 160, 161
Freeman, Charles 114n4
Freemark Abbey (winery) (formerly Lombardi) 79, 96, 114n4
French Paradox 188n21
French Revolution 38–9
Friszolowski, Mark *xii*
fruit-and-nut crops 18, 19, 68, 74, 84, 98

G
Gallagher, Patricia 103
Gallo, E & J 86–7, 101, 164, 182
Gamay 82
Garonne river (Bordeaux) 35
Gauls 32
gender 27, 65, 66, 94, 100
Georgia (Europe) 26*illus*, 27, 28
Gewürztraminer 90
GI Partners 164, 182
Girard winery (later Vintage Wine Estates (VWE)) 165
Gislason, Nick 176
Glen Ellen (distribution network, Sonoma) 60
gold 49
Gomberg Frederikson & Associates 92n27
Gomberg, Louis 85, 86, 92n27, 97
Gomberg Report 85
Grace Family Vineyards 182
Grace, Richard and Ann 127

grand cru 39
Granholm v Heal (2005) 178–80
grape-growing *see* viticulture/viniculture
grapes and grape juice concentrates 68, 69, 71n30, 81, 84, 98, 165, 177, 178
Graves 39
gravity-flow fementation techniques 57
Great Depression 68, 72, 75
Grenache 57
Greystone Cellars (distribution network, Napa valley) 60
Grgich Hills (winery) 105
Grgich, Mike 87, 88, 95, 105, 112, 113n1
Groth (winery) 111
Gundlach, Jacob 52

H
Hanseatic traders 35
Haraszthy, Agostin 50, 52
Harlan, Bill *xii*, 124–5, 141n14, 141n19, 149, 165–6
Harlan Estate 124–6, 149, 165–6, 169n42
Hermosa Vineyards (later To Kalon Wine Company) 170
Heublein 89, 107, 129, 188n19
high-quality wines 8, 109, 116, 119–20, 121, 153, 159, 173, 186
  historical background 26, 32, 33, 34
  post-Prohibition 76, 77, 81, 82–3, 84
  reputation building in C19th 39, 40, 43–4, 50, 52, 58, 59, 64, 69
high-yield varietals 40, 50, 67, 69, 86
Hilgard, Eugene 55–7
Hills, Austin 105
hillside vineyards 52, 57, 122, 124, 127, 146–7, 152, 161, 162
Hills Vineyard 105

home winemaking 67, 74
Honig (winery) 111
Hospices-de-Beaune Wine Auction (Burgundy) 175
Howell Mountain 89, 127, 132

I
immigrants and immigration 5, 49, 52, 59, 65, 85, 101, 114n4, 119, 186
independently owned wine shops 3
industrial alcohol 42
Inglenook (winery) 89, 92n14, 92n26, 94, 126, 141n19, 188n19
  high-quality wines 57, 58, 64, 75, 77, 79
Inglis, David 24
innovation *see* technological/scientific innovation
Insignia (wine brand) 126
Institut national del'origine et de la qualité (France) 132
intra-firm networking 22
Italy 1–2, 8, 25, 32, 44, 114n4, 132, 148, 164

J
Jaeger, William 96
Johnson, Chuck *xii*
Jurade of Bordeaux 36

K
Kedge Business School *xii*
Kenward, Tor *xii*
Kingsburg 79
Kirin Brewery 129
Knights Valley 88
Kohler & Frohling 49, 54, 60
Korbel 92n26, 97
Kramer, Matt 120
Kroenke, Stan 149, 182
Krug, Charles 52, 54, 55, 59, 79
kvevri (storage vessels) 26*illus*, 27, 28

L
labelling *see* brand identity
labour force 21, 40, 46, 49, 80, 152, 167, 171, 174, 184, 186
L'Académie du Vin (Paris) 102–3
Lafitte Rothschild (Bordeaux) 9
Lafitte (winery) 114n25
Lail, Robin 141n19
land-use practices 20, 21, 98–9, 129–31, 133–5, 161–3, 171, 187
Lapsley, James 72, 81–2, 83, 85, 100, 101
Larkmead (winery) 75, 92n14
La Tour (Bordeaux) 9
Laube, James 79, 125, 150, 158
Levy, Bob 124
Lombardi (winery, later Freemark Abbey) 79, 114n4
London Gazette 9
Los Angeles 49, 60
Los Gatos 61, 92n26
luxury wines 151, 155, 164, 166, 174, 175, 178, 182, 184
LVMH 182

M
McIntyre, Hamden 57
Maiden, The (Harlen Estate) 125
Malvoisie 52
Margaux (winery) 114n25
market organization 12–16, 22, 24, 25
Markham Winery 129
Marks, Denton 15–16
Marshall, Alfred 14
Martha's Vineyard 84
Martini, Louis M. 79, 80, 87, 88, 92n14, 164
mass-produced wines 2, 3, 39, 40, 42, 43, 70, 95, 109, 110, 177
historical background 26, 30, 31, 32, 35
ownership 181, 182
post-Prohibition 72, 73, 74, 75, 77, 78
post-Second World War 79–82, 84, 86–7, 90, 100, 101–2, 112, 118–19, 133
Mathews, John A. 13–14
May, Tom 84
Meadowood Resort 124
mechanical refrigeration 79
Medoc ('Left Bank', Bordeaux) 37, 39, 62, 111
Merlot 107, 111, 124, 136, 146, 149, 175, 176
Merryvale Winery (formerly Sunny St Helena) 124, 125, 141n14
Meursault Charmes 104
Meyer, Brother Justin 95
Middle Ages 5, 8, 25, 32–5, 43, 46
Middle East 27
Mihaly Winery 129
mildew 40
Miller, Dianne and Ron 126–7
Ministry of Agriculture (France) 109
missionaries *see* Franciscan missionaries
"Mission grapes" 46–7, 49, 50, 52, 53, 56, 57, 58, 68–9, 82
Mission San Diego de Alcala 46
Mission San Francisco Solano (Sonoma) 46, 47, 48
Moët-Hennessy 97, 108
monastic orders 5, 8, 25, 32–4, 38, 39, 43
Mondavi family
 Cesar 79, 88, 141n14
 Margit 175
 Michael 109
 Peter 79, 87
 Robert 20, 79, 87–8, 92n14, 95, 109–11, 112, 124, 128, 164, 175
 Timothy 109, 110, 164
Mondavi (winery, Rutherford) 87–8, 92n14, 95, 141n14, 141n31, 163–4, 170
Mont La Salle (winery) 75
Moueix, Christian 123–4
Mourvedre 57
Mouton Cadet (winery) 110

Index

Mt. Veeder (winery) 164
Muscat 52

N
Napa County Board of Supervisors 134, 162
Napa County Planning Commission 99, 162
Napa Creek (winery) 133, 159
Napa Ridge (brand) 159
Napa River 161, 162
Napa Valley 18, 53, 68–70, 76*illus*, 82*illus*, 92n26, 169n35, 186–7
  appellations and sub-appellations *ix*
  definition of 'Napa Wine' 159–61
  emergence of quality wine 51–7, 59
  hierarchical fragmentation 157, 158, 167, 174, 187
  Prohibition and collapse of wine industry 65–8
  resurgence of 61–5, 172–5
  signature grape 122, 131, 148, 149, 164, 172, 175–6, 187
  suburbization and 98–9, 113, 161
  winery interview details 190–2
Napa Valley Cooperative Winery 84, 92n14, 95
Napa Valley Enological Research Laboratory (1945) 83
Napa Valley Grape Growers' Association (NVGGA) 102, 106, 108, 146, 148
Napa Valley Growers Association 134
Napa Valley Reserve (private winegrowing estate) 165–6
Napa Valley Vintners Association (NVVA) 80–1, 86, 102, 106, 134, 139, 145, 160, 175, 185
Napa Valley Wine Technical Group (NVWTG) (1947) 83, 122, 126, 138

Napa Wine Company 185, 188n19
Napoleon III, Emperor 37
négociants see wine merchants
Netherlands 25, 35
Newton Vineyard 182
New York 49
New Zealand wines 2
Niebaum, Gustav 58, 79
North Carolina *x*, *xii*
Northern Europe 39
Nouvelle Aquitaine (Bordeaux) 35
NVGGA *see* Napa Valley Grape Growers' Association
NVVA *see* Napa Valley Vintners Association
NVWTG *see* Napa Valley Wine Technical Group

O
Oak Knoll 50
Oakville 62, 80, 84, 111, 124, 133, 148, 149
Old World producers and connoisseurs *xi*, 18, 172
  competition from New World wines 102, 104, 106, 110, 111, 116, 186, 187
  synthesizing of knowledge 56, 120, 124, 136, 139
  ultimate benchmark 20, 22, 25, 59, 70, 72, 85, 149
Opus One (winery) 20, 110–11, 151, 163, 170
Ordinance #947 (seventy-five percent rule) (1990) 135, 160, 163
organic agricultural practices 161, 162
Osborne, J.W. 49–50
Ouvrard, Stephane *xii*, 38
overproduction 59–61, 84
ownership 20, 45n26, 53, 55, 116, 143–5, 175, 181–2, 190–2
  France 36, 37, 38, 39, 43
  1960s/1970s 95, 96, 98, 99, 112, 113

pre-/post-Second World War 74, 75, 76, 79, 80, 88, 90–1
quality wine 120, 121–9, 134, 135, 139
see also corporate ventures and investment

P
Palomino 82
Panama-Pacific International Exhibition (PPIE) (1915) 64
Paris Tasting (1976) 20, 102–6, 109, 111–13, 116, 123, 138, 139, 164
Paris Universal Exhibition (1855) 37
Parker, Robert 10, 123, 147, 149, 150, 154, 158, 164
partnerships 96, 105, 106–11, 114n4, 123, 164
Pasich, Leland 95
Penrose, Edith 145
Pepys, Samuel 9
pesticide use 161
Petaluma Rancho 47
Peterson Barrett, Heidi 148–9
Peterson, Richard 148
Petite Sirah 20, 67, 74, 82, 89, 98, 112, 131, 146
Petit Verdot 107, 124, 136, 146, 149
Phelps, Joseph 126
Phillips, Jean 148–9
Phillips, Rod 24–5, 32, 34, 36
phylloxera 40–1, 108, 114n4, 124, 127, 146, 170
    AxR#1 (phylloxera-resistant rootstock) 117–18, 130–1
    early years of Californian wine 55, 59, 62, 68
Pillsbury 89
Pinney, Thomas 74
Pinot Blanc 97
Pinot Meunier 97
Pinot Noir 34, 58, 68, 118, 146, 176–7
    Pre-/post-Second World War 82, 89, 90
    Sixties/Seventies 94, 97, 98

Place de Bordeaux (France) 43, 151
Pomerol 123, 124
Portugal 4, 25, 44
PPIE *see* Panama-Pacific International Exhibition
premier cru status (first growth) 39, 109
premiumization 30, 35, 61, 62, 67, 68, 97, 116, 139, 178
    cult/luxury wines 143, 148, 150–1
    growth in 1970s 108, 111, 112
    pre-/post-Second World War 81, 85
Premium Wine Producers in California Association (PWPCA) 86
prices and pricing power *xi*, 9, 15, 16, 20, 75, 80, 90, 100–1, 166
    economic downturn and 59, 60–1, 62, 63
    historical background 27, 30–1, 35–6, 37, 39, 42, 45n26
Pritchard Hill 127, 149
production 6, 7, 16, 18, 19, 24, 27, 100, 176–8
    California in C19th/C20th 52–3, 54, 61
    historical background 30–2, 35
    monastic orders 5, 8, 25, 32–3, 34, 43
    post-Second World War 83–6, 98, 144
    small-batch (cult wines) 148–52, 156
profit maximization 81, 96, 111, 113, 136, 144, 145, 174
Prohibition (United States) *xi*, 17, 19, 114n4, 130, 184
    collapse of Californian wine industry 64, 65–8, 70, 170
    post-Prohibition decades 72–5, 77–8, 90, 94, 95, 116, 121, 129
Promontory (Harlan Estate) 125
prune crops 71n30 74, 84, 98

PWPCA *see* Premium Wine Producers in California Association

Q
quality 4, 8, 15, 16, 18, 28, 30–1, 32, 57
  Cabernet Sauvignon and 146–8
  definition of 9–12
  quality versus quantity debate 17, 53–4, 56, 69, 81, 100
  systematizing of 25, 33–7, *38*, 39–42
quantitative scores 10

R
Raffaldini, Ray *xii*
Ray, J.W. *xii*
Raymond Winery 129, 162
Reata (winery) 162
regional/geographical identity 9, 14, 15, 21, 36, 106, 121
  early years of California wine 55–6, 59, 61, 69, 70
  land-use practices 132–3, 134, 172
  Napa Style 135–8, 167
regulatory environment 9, 15, 16, 20, 21, 31–2, 50–1
  classification and 132–3
  land-use practices 161–3
  Prohibition (United States) 17, 19, 65–8
  systematizing of quality 25, 33–7, *38*, 39–42, 187
religion and spirituality 5, 7, 17
  Ancient World 24, 25, 27, 29, 30, 32–3
  Franciscan missionaries 46–7, 48*illus*, 68
  monastic production 5, 8, 25, 32–3, 34, 38, 39, 46
  sacramental wine 8, 32, 34, 67, 69, 114n4
Remaud, Hervé *xii*
"Report on Grapes and Wines of California" (Haraszthy) 50
reputation building *xi*, 21, 22, 91, 98, 111, 156, 167, 176
  early years of California wine 53, 54, 55, 57, 59, 61, 63, 64
  historical background 36, 37, 39, 41
  post-Prohibition 70, 73, 75, 85, 86, 116, 117, 121, 125, 132, 139, 144
  promotion during 1970s 102–6
Rhone Valley 1
Ridgeway, Cecilia 119
Riesling 52, 57, 58, 90, 105, 126, 146
Riorda, Jack 141n14
rituals and celebratory feasts 5, 7, 8, 17, 24, 28, 29, 30
Robbins, Michael 89
Robert Mondavi Corporation 164
Robinson, Jancis 137, 147
Rolland, Michel 149
Roman Empire 7, 8, 17, 24, 25, 28, 29, 30, 31–2
Roney, Rat 165
rosé style wines 45n20, 52, 101
Rothschild, Baron Phillippe 20, 109–111, 114n25
Rudd, Leslie 165
Rupestris St George rootstocks 117–18, 130
Rutherford Vintners 133, 159

S
sacramental wine 8, 32, 34, 67, 68, 69, 114n4
St Clement (winery) 129
St Emilion ('Right Bank', Bordeaux) 39
St Helena Cooperative Winery 84, 95
St. Helena Vinicultural club 80
St Helena Winegrowers Association 54–5
St Helena (winery) 52, 53, 79, 87, 124

sales 33, 58, 59, 61, 152–5, 178–81, 184
Salmina, Elmer 92n14
San Francisco 75, 80, 84, 98, 182
early years of California wine 46, 47, 49, 50, 53, 57, 59, 60, 64, 69
San Gabriel Mission 47
San Juan Capistrano Mission 46–7
Sauternes 37
Sauvignon Blanc 2, 62, 146, 176
Schlatter, Jack 141n14
Schrader Cellars 149
Schram, Jacob 52
Screaming Eagle (cult wine) 148–9, 176, 182
Scruton, Roger 11
second labels 125, 177
Second World War 17, 72, 81
Sémillon 62
seventy-five percent rule (Ordinance #947) (1990) 135, 160, 163
Shafer, Doug *xii*, 123
Shafer, John 121–3
Shafer Vineyards 121–3
Shafer (winery) 111
Shelton, Tom *x*
Silverado Tavern 126
Silverado Trail (Stag's Leap District) 122, 126, 132, 159
Silverado Vineyards 126–7, 141n31
Silver Oaks Cellars 95
Sionneau, Lucien 110
Sirah/Syrah 126, 175–6, 177
small wineries 53, 144, 152, 171, 172, 176, 178–80, 181, 182–5
sociability 5–6, 11, 24
soil conditions 9, 17, 25, 30, 34, 106, 117, 120, 124, 132
    early years of California wine 50, 51, 52, 56, 62, 69, 170
    erosion 161, 162
Sonoma County 118, 176
    early years of California wine 46, 47, 48, 50, 51, 52, 53–4, 55, 56, 59, 60, 61
Sonoma Index 53–4

Sonoma-Napa Horticultural Society (1859) 50
South of France 40, 105
Souverain Cellars 87, 89
Spain 25, 44, 47, 132
sparkling wines 97
Spotteswode (winery) 111
Spring Mountain (vineyard) 89, 133
Spurrier, Steven 102–3
Stag's Leap Wine Cellars 104, 105, 132, 141n31, 164
State Fair in California (1940) 80
state legislation 50–1
State Viticultural Convention (1888) (San Francisco) 57
Stelling Jr., Martin 80
Ste. Michelle Wine Estates 164
Sterling (winery) 95
Stewart, Lee 87, 89
storage of wine 26*illus*, 27, 28, 53
Stralla, Louis 92n14
Sullivan, Charles 47, 56, 58, 68, 79, 147, 166
Sunny St Helena (winery, later Merryvale Winery) 79, 87, 124, 141n14
super-value wines 118–19, 133
supply and demand 12–13, 43, 52, 59–60, 94, 172, 176
Supreme Court (United States) 159, 160, 178–9
sustainability *see* environmental issues
Sustainable Wine Growing Program (Wine Institute) 161
Swanson Vineyards 165
sweet wines 5, 28, 35, 52, 63, 74, 80, 83, 136
Sylvaner 52
symposiums (*symposion*, Ancient Greek) 4, 30

T
Taber, George 103, 104, 113n1
table wines 57, 67, 116, 119
    emergence of wine culture 101,

102, 103
  pre-/post-Second World War 73,
    74, 75, 80, 83
Taplin, Ian 38
tariffs 55
Tari, Pierre 104
taste preferences 36, 91
  subjectivity of 1, 2, 5, 9, 10, 11,
    12, 28
tasting rooms 96, 105, 123, 135, 171,
  179, 180–1, 183, 184, 185
taxation 50, 55, 85, 96, 131
Taylor California Cellars 101
Tchelistcheff, André 73, 78–9, 80,
  83, 87, 113, 122
technological/scientific innovation
  25, 28, 78, 83, 87, 116, 117, 120,
  144
  California in C19th/C20th 55,
    56, 57, 58, 62, 64
  collective organizational
    learning 12, 13–14, 15, 17,
    18, 20, 21, 22
  current trends 172, 173–4, 186
  disease 129–31, 146
  France 35, 39, 42, 43–4
  growth in 1970s 105, 109, 112
temperance movement 29, 65–6
terroir 8, 18, 36, 37, 42, 43, 55, 106,
  117, 124, 125, 138, 152, 157
Thach, Liz 155
Time 103
Todd, Cain 11
To Kalon (formerly Hermosa
  Vineyards, Oakvil) 62, 80, 88,
  99, 108, 109, 170
tourism 21, 99, 133–4, 171, 180
trade organizations *see* advocacy
  organizations
Transactions (California State
  Agricultural Society) 50
transportation 15, 28, 35, 39, 40, 77
Trefethren (winery, formerly
  Eshcol) 57, 58*illus*, 68, 95
"trophy wineries" 143
TSG Consumer Partners 182

Tubbs, Alfred 95
Tuscany 4, 32
"two buck chuck" phenomenon 119,
  160
Tychson, Josephine 114n4

U
UC Berkeley 76–7, 78
UC Davis 73, 76–7, 78, 83, 84, 90,
  94, 105, 122, 130, 138, 144
ultra-premium wines 111, 123, 138,
  151, 155, 163, 164
  production and distribution 173,
    174, 178, 182, 183, 185, 186
United Vintners 89, 107
University of California (UC) 55, 75,
  76, 77
university research 55–6, 61, 113,
  117–18, 138, 155–6
  phylloxera 129, 130, 131
  pre-/post-Second World War 73,
    75, 76–8, 83, 88, 90, 94
USDA *see* Department of
  Agriculture

V
Vallejo, Mariano 47, 48
varietal identity 16, 17
Viader, Delia *xii*, 127
vignerons 6, 9, 41
vignobles de marchands (merchant
  estates) 39
vignobles de paysans 38–9
vineyard management 106–8, 110
Vinifera Development Corporation
  107
Vintage Wine Estates (VWE)
  (formerly Girard winery) 165
viticulture/viniculture (grape
  growing) 16, 17, 19, 72, 100,
  117–18, 131, 146–7, 173
  Ancient World 24, 25, 27, 30,
    31, 32
  California in C19th/C20th 50,
    51, 52, 53, 56, 61–3, 67
  Franciscan missionaries 46–7

Middle Ages 33–4, 39, 41–2
premium varietal plantings in
 1960s 89–90
Volstead Act (prohibition) (1919) 66

W
Ward, Alan *xii*
Watson, William 58
Weaver, Don 126
Weis, Russ *xii*
Wells Fargo Bank 96
Wetmore, Charles 57
Whitehall Lane (winery) 129
White Riesling 57, 58
WI *see* Wine Institute
William Hill Estates 164
Williams, Craig 126
wine
  approach to selecting/buying 2–3
  destabilizing effects of 29, 31
  effect of environment and
   context 10, 11–12
  historical background 4, 6–8, 25,
   26, 27–42
  luxury product 3–4, 8
  origins of 25, *26*, 27–30
  recommendations of 1–2
  as a social/cultural product 4–6,
   9, 24–6, 32, 100, 101, 116,
   119–21, 122
Wine Access 176, 181
Wine Advocate 147
wine brokers (*courtiers*) 35, 36, 43

wine clubs 152, 179, 180, 181, 1
wine collectors 154–5
Wine.com 181
wine critics 10, 21, 118, 123, 125,
 137, 143, 147, 149, 150, 151,
 154, 155, 163–4, 166, 173
wine estates 35, 37–9
Wine Institute (WI) 75–6, 161–2
wine merchants and traders
 (*négociants*) 35, 43, 49, 53–4,
 55, 59–61
Winery Definition Ordinance (1990)
 162, 171
Wine Spectator 10, 79, 143, 147,
 150, 153, 154, 158
Wines and Vines 90, 93n29, 130
Winiarski, Warren 87, 105, 112, 164
Winkler, Albert 73, 77, 83, 84
Woodbridge brand 95, 109, 110, 163

Y
Yeung, Peter 155
Yount, Eliza 170
Yount, George 48, 170
Yountville 48, 49, 97, 99

Z
Zhao, Wei 16
Zinfandel (formerly Zinfindal) 20,
 74, 82, 89, 98, 112, 118, 131,
 146, 176
 popularity in C19th 50, 52, 57,
  58, 62, 67